Private Science

The Chemical Sciences in Society Series

Arnold Thackray, Editor

Sponsored by

Private Science

Biotechnology and the Rise of the Molecular Sciences

edited by Arnold Thackray

PENN

University of Pennsylvania Press

Philadelphia

Printed in the United States of America on acid-free paper

10 9 8 7 6 5 4 3 2 1

Published by
University of Pennsylvania Press
Philadelphia, Pennsylvania 19104–4011

Library of Congress Cataloging-in-Publication Data
Private science : biotechnology and the rise of the molecular sciences / edited by Arnold Thackray.
p. cm. — (The Chemical sciences in society series)
Includes index.
ISBN 0-8122-3428-6 (alk. paper)
1. Biotechnology—History. 2. Biotechnology—Political aspects.
I. Thackray, Arnold, 1939– II. Series.
TP248.18.P75 1998
306.4′6—dc21 97-31823
CIP

Contents

Part III: The Molecular Workplace

Introduction

Arnold Thackray

More than half a century has passed since the events that signaled the birth of modern biotechnology. Initial utopian hopes and Faustian nightmares seem equally overblown. Today, age-old dreams of elixirs and magical potions increasingly yield to the specialized, sanitized skills of biomolecular science. One indicator of the distance traveled since the 1944 announcement that DNA was the messenger of life may be seen in today's reality of the existence of the flourishing Biotechnology Industry Organization, which represents over 250 public companies and perhaps another 1,000 privately held corporations. Another indicator lies in the federal government's commitment to U.S. strength in this field, as mediated primarily through the National Institutes of Health and the massive, fifteen-year "big science" program originally known as the Human Genome Initiative, now the Human Genome Project.

With the ending of the Cold War, the military-industrial complex in its old, familiar sense is passing from the scene. New configurations of federal power, private initiative, academic research, and corporate interest in science-based technology are moving to center stage. Chief among them is that fuzzy complex of actors and actions gathered under the label of "biotechnology."

Identifying profound revolutions in human thought and behavior and assimilating them into our patterns of understanding can take many decades. The now-familiar Scientific Revolution of the seventeenth century, and its close cousin the Industrial Revolution of the late eighteenth and early nineteenth centuries, became staple items in historical discourse only when Western civilization took stock of its common modernity, in the years following England's Great Exhibition of 1851, the U.S. Civil War in the 1860s, and Germany's unification in 1870.

Any thoughtful observer or amateur historian is aware that the Scientific and Industrial Revolutions spawned further, major transformations in the patterns of human life. If the dominance of factory over farm was the essence of the Industrial Revolution, today both factory and farm are being superseded by the information and service professions. There is as yet no settled consensus on what to call this transformation to the knowledge-based society that seems destined to set the agenda for the opening days of the new millennium. It is apparent that information, primarily in electronic and computational form, holds one key. And it is increasingly apparent that the promise of a full and healthy life span for all citizens, mediated through the revolution in biotechnology and the life sciences, holds another key. Thus, slowly and painfully, historical curiosity is being aroused by and historical attention is beginning to be focused on biotechnology and the "biomolecular revolution" as one key component in defining the post-modern world.

If one asks where the Scientific Revolution or the Industrial Revolution began or ended, the answers of historians will illustrate a general consensus. At the same time individual historians will vary widely in their interpretations of how the elements — the individuals, the countries, the concepts, the events, the people, and the chronologies — are woven together. This variation of interpretation is all the more apparent with the Biomolecular Revolution we are still experiencing! Because that revolution is in full swing, its analysis has thus far depended heavily on the efforts of journalists and of the more self-aware participants. Books such as *Gene Dreams* and *The Gene Wars* display the power and the limitations of such contemporary reporting. They also form the backdrop against which more academic and sociological works such as *The Code of Codes* or *Making PCR* are now beginning to set the stage for a fuller appreciation.

Private Science is thus offered as a contribution to the new discourse on the structure and significance of biotechnology and the Biomolecular Revolution. Its chapters are written by noted historical and sociological scholars who had the additional advantage of meeting together in conference to discuss and critique each other's drafts, and who were thus able to utilize one another's insights in preparing their essays for this volume. While each essay stands on its own, their cumulative significance lies in the way they point up wider, salient themes necessary to our rapidly growing understanding of biotechnology and the Biomolecular Revolution.

The very choice of *Private Science* as the title for this book signals the ambiguities and complexities that confront the student of biotechnology. Is molecular biology as an academic pursuit, a "public good," compromised in any way by the hopes of private gain that attend its commercial sponsorship and exploitation? Indeed, what are we to understand by private and public in the domain of science? Turning our usual notions on their head, Alberto Cambrosio and Peter Keating argue that the public nature of biotechnology must be constructed from the private science of local networks. Stephen Hilgartner shows how academic researchers commandeer "public" resources for "private," competitive motives. Thomas Gieryn shows how complex is the design and symbolism of public and private spaces for biomolecular research. *Private Science* examines such issues in the contexts of political economy and comparative history and maintains a close focus on core studies of scientific practice. These wider themes pervade the individual essays.

These twelve essays are grouped in three sections, which proceed from prehistory to politics to practice. Like *science* or *industry*, the terms *biomolecular sciences* and *biotechnology* itself are words with long histories and ramified usages. It is therefore appropriate that the very first article, by Robert Bud, deals eloquently and succinctly with the long-term history of biotechnology. As the overall heading to the first section of that essay, "From Brewing to Biotech," reveals, brewing—and food technologies—form the time-honored context against which more modern usages and new inventions for the word *biotechnology* must be understood.

If brewing was the technology that underpinned the old, biochemical biotechnology, genetics is the science that informs the new, molecular biotechnology. Much of the rest of this book is about the continuities and the discontinuities in the shift from the biochemical biotechnology of brewing and fermenting to the molecular biotechnology of DNA, cloning, and gene splicing. Lily Kay, for example, focuses on that "molecular vision of life" which, gathering force from the 1930s to the 1960s, helped shape the public expectations and media hype that greeted the new molecular biotechnology of the 1970s. And in a third stage-setting essay Angela Creager analyzes how even the "new" molecular biotechnology of the 1970s and 1980s owed much to earlier *social* techniques for the "corporatization" of academic science. These techniques were pioneered in the pressure cooker of World War II and perfected in the late 1940s and the 1950s. Here the focus is not the biochemical art of the brewery but the academic biochemist's science-based sequences for blood separation. Much like Kay, Creager argues for the importance of "the longer-term history of entrepreneurial technology within 'basic' molecular biology and biochemistry." She convincingly establishes that many aspects of laboratory organization, government involvement, and patent policy that would later become crucial to the "new" biotechnology were elaborated by Edwin Cohn at Harvard in his plasma fractionation project.

The essays in Part I thus provide admonitory contexts, warning us not to overlook the historical roots or precedents of the new rhetorics and uses for the biomolecular sciences and biotechnology. Nonetheless, *something* was very new about the constellation of forces at play in the 1970s and 1980s. Venture capital, legal policy, national definitions of appropriate "reasons of state," scientific concepts, and available technique all came together to create a situation in which—especially in the United States—the scientific and industrial momentum of biotechnology itself was in peculiarly fruitful juxtaposition with the socioeconomic and political climate. As Susan Wright observes in her study of molecular politics, not only the socioeconomic climate but the interests of the technical elite itself were reshaped in the early 1980s. "Momentum" was highly important, for "Had genetic engineering been merely a powerful research tool—had it not also provided the basis for a new form of industrial production—it seems likely that the scientific community's rationales for deregulation [of genetic engineering] would have met with far greater skepticism."

The essays in Part II explore legal, political, economic, and symbolic aspects of this new constellation of forces. "Macromolecular politics" is their common theme. Daniel Kevles describes in detail one crucial discontinuity (or continuity?), namely the political economy of the patenting of living organisms, from Ananda Chakrabarty's oil-consuming bacterium to the so-called Harvard mouse. Kevles's highly specific account is nicely complemented by Susan Wright's comparative essay. Using the approaches of a post-pluralist history, Wright powerfully contrasts the different course of the debate over genetic engineering in Britain and the United States during 1972–1982. Her emphasis on the global and comparative aspects of the new molecular politics finds a useful foil in Herbert Gottweis's detailed examination of the political economy of British biotechnology. Gottweis's subtle and persuasive analysis links theories of economic growth, trends in political ideology, and the unintended consequences of government policy to reveal why British and American biotechnology took such different paths.

"Political economy" occurs in the titles, or informs the contents, of what Kevles, Wright, and Gottweis wish to establish about the new molecular biotechnology of the 1970s and 1980s. The new biotechnology differed from the older biochemical biotechnology of brewing and fermentation not least in its lack of an immediate and obvious commercial viability. Indeed, a new company's "burn rate" (i.e., the rate at which it consumes its financial capital, en route to its distant—and hopefully, profitable—landing) has become one indicator of its ambition. Appropriately, Martin Kenney explores the crucial process by which the new molecular biotechnology came to be conceptualized as a subject worthy of its own economic space. His Schumpeterian model of this new economic space provides an important analytic framework for making sense of twenty years of rapid growth in this industrial sector. His analysis thus builds on, and usefully illuminates, the discussions of patents and of molecular politics in the earlier essays in this section. Finally, in the last essay in this section, Sheldon Krimsky focuses explicitly on the continuing struggle over the cultural and symbolic meaning of biotechnology. He shows how various actors have argued for the natural or unnatural character of biotechnology, stressed its contributions to greater (or lesser) biodiversity, shown it as friendly (or unfriendly) to the environment, and seen it as offering greater (or lesser) control over nature.

He thus captures the protean nature of those macromolecular politics that shape the character and context of all modern technoscience.

Part III deals not in macropolitics but in the more local and detailed actions of biomolecular science as it is lived in the laboratory, the databank, the genome project, and—glancing back at the larger world of symbols—the design of its living spaces.

Alberto Cambrosio and Peter Keating's counterintuitive approach shows how the tools and techniques of the biomolecular sciences are themselves complex social constructions, and how the traditional distinctions between "public," or academic, and "private," or commercialized, science are themselves increasingly problematic. This same concern with the problematic nature of the distinction between public and private science pervades Stephen Hilgartner's exploration of policy on data access in genome research. Genome research is also the focus of Michael Fortun's examination of big science as "fast science," which exposes speed as a matter of social and political choice. And concluding our volume, Thomas Gieryn's deconstruction of the design of spaces where biotechnology is pursued offers an intellectual tour de force. Decoding the symbolisms of space and architectural construction, he shows how they relate to biotechnology as simultaneously a private and public science.

The essays in this volume can only hint at the growing, variegated riches available for exploration in the modern bimolecular sciences. That is as it should be, for who can hope to offer the last word on a revolution—especially in its midst!

Part I
From Brewing to Biotech

Molecular Biology and the Long-Term History of Biotechnology

Robert Bud

Biotechnology: Repute and Content

> What is reported of men, whether it be true or false, may play as large a part in their lives, and above all in their destiny, as the things they do.
> —Victor Hugo, *Les Misérables*[1]

For technologies as for people, Victor Hugo's dictum suggests that repute is no simple by-product of actions. The reputation of biotechnology is based almost entirely not on what has happened but on what is to come: leaders across the industrialized world, science policy analysts, and industrialists are agreed that a currently modest industry will soon be enormous and a major contributor to industrial renewal. The general public has, in contrast, on occasion exhibited a terror of the new technology that exceeds the implicated threats of any particular applications. As Sheldon Krimsky shows in this volume, in the United States, Germany, and elsewhere the engineering of individual organisms is characterized as a prelude to the destruction of nature and the manipulation of the human species. The "public understanding of biotechnology" is still a major concern for those promoting it on both sides of the Atlantic.[2]

Whatever our own judgment of probabilities, we should not define any one of these prophetic views as "correct." Skeptics have dismissed expectations as exaggerated. The industry has developed neither as quickly nor as independently as first hoped. Many of the pioneering companies, rather than becoming the IBMs of the future, have been taken over by older organizations with greater experience of the market. During the later 1970s specific hopes were concentrated on manufacturing first insulin and then interferon: solutions to the key diseases of diabetes and cancer seemed within reach. Yet even when insulin and interferon did not live up to their promise, hopes for biotechnology as a whole survived. Market surveys anticipating enormous turnovers by 2000, such as T. A. Sheets's prediction that sales in biotechnology would approach $64.8 billion, were strongly criticized by product experts and apparently unsupported by the therapeutic market, yet the optimists persist.[3] In 1997 there is still a dispute as to whether a major new industrial sector is on the verge of maturing or whether new techniques have merely been assimilated by existing industries.[4] Strangely, perhaps, much more effort has been invested in understanding the cultural bases of fear than in analyzing the countervailing hopes.[5]

Commentators express the long-standing fear of breaching the divide between the quick and the dead in terms much like those Mary Shelley used in *Frankenstein.* They less often note that the description of biotechnology as the technology of the future, though not as deeply grounded, is at least as old as the twentieth century. Whether such expectations are reasonable or inflated, their careful articulation, wide acceptance, and translation from one technique to another have made them important. They provide a lens through which calls for more resources and reports of success and failure have been seen. It has been so much part of our culture that this constant willingness to believe in biotechnology is easily overlooked.

Extravagant forecasts for other new technologies such as plastics, space, and communications each have their own traditions.[6] While such forecasts are clearly not free-floating phenomena independent of how well the various techniques perform, but are extrapolated from experience, predictions of biotechnology's future are not determined merely by how well laboratory techniques actually work. To borrow a term used by Charles Rosenberg in describing the "fashioning of explanatory and classificatory schemes of particular diseases," historic wishes for a benevolent biotechnology can be said to have helped frame the perceptions of the 1970s.[7] I argue here that, irrespective of the justice behind the hopes, they were first expressed in terms of a deeply entrenched model of biotechnology, with an urgency that was a response to threat.

Even the word *biotechnology*, whose combination of "bio" and "technology" so evocatively encapsulates those hopes, is itself not brand new.[8] Widely used during the 1970s, particularly in Europe, the word was entered in major German dictionaries half a century earlier, as *Biotechnologie*, and was commonly applied to the uses of microorganisms and the processes of fermentation.[9] It now connotes the application of very recent biological techniques, and as recently as 1982 the geneticist Zsolt Harsanyi declared, "Biotechnology is a neologism."[10] As the author of *Impacts of Applied Genetics*, the first study of the subject conducted by Congress's Office of Technology Assessment, Harsanyi was well qualified to comment, and, as an employee of the investment house of E. F. Hutton, he knew that the firm had taken out a copyright on the word in 1979. Shortly after, biotechnology's image as a distinct brand-new industrial base made it appear a golden investment opportunity for the whole of Wall Street. When the first biotechnology company shares were launched in 1980, they rose faster than any other stock in the history of the New York capital market.

Beyond its meaning as the occasion for a stock market bonanza, biotechnology seemed to be a potential savior of the American economy and a validation of the nation's global dominance of biomedical research. After all, Americans were credited with most of the discoveries lumped under that rubric: the technique of recombinant DNA, invented by Stanley Cohen and Herbert Boyer in 1973, and earlier the double helix itself, in which James Watson played such a visible part (even if he was working in foreign lands with foreign coworkers at the time). Congressional hearings repeatedly emphasized the way in which the practical prospects for the new technology indicated recent U.S. investment in basic biomedical sciences. A typical example is Chairman John D. Dingell's remarks to the House Committee on Energy and Commerce: "Both the new scientific understanding and the potential for solving society's problems make these advances revolutionary. These potentially extraordinary benefits are the fruits of years of basic research funded primarily by the U.S. Government through the National Institutes of Health and the National Science Foundation."[11] Historical comment has tended to echo this perspective, equating the history of biotechnology with the recent emergence of applications of molecular genetics.

By the early 1980s, to judge by reporting in the general press, no connection was made between the apparently boring descendant of brewing, the "old" biotechnology, and the exciting "new" genetic technology. Yet one can identify many continuities. A small amount of engineered DNA does not an industry make. Organisms must be chosen, grown, harvested, and processed, and products must still be tested and sold. By the late 1980s the established pharmaceutical and agrochemical companies still dominated the biotechnology industry.

My argument goes further: I suggest that the molecular biologists who promoted the new biotechnology appropriated not just old techniques but also the old visions, well established and well known within a limited circle, to promote a new paradigm to a much broader world. Awareness of this process can help us better interpret the enduring nature of a technology that otherwise all too often appears to have materialized miraculously in the research laboratory at the end of this century to rescue a dying industrial system.

The Biotechnic Age

The image of an "age" based on a new industry has long dominated the twentieth century. The model of the stone, bronze, and iron ages—first developed by C. J. Thompsen for the 1836 Copenhagen Museum Guide—stimulated thoughts of modern analogues. By the end of the nineteenth century, in an era of evolutionary thinking, the civilization of coal and iron pioneered in Industrial Revolution Britain was itself clearly being superseded.[12] New technologies and new national styles were asserting themselves. The application of science, particularly chemistry and physics, seemed to characterize a new basis for industry. Next in the hierarchy of sciences was biology. As early as 1892, the American geologist Clarence King was proposing: "This is the age of energy; next will be the age of biology."[13] He made this prophetic judgment in the name of a new scientific world that he believed could boast two radical new ideas: the conservation of energy and the theory of evolution. Hitherto, he felt, only the first had been properly applied; now it was time for the second of these great ideas to be turned to practice.

In the years before World War I, engineering, rather than the application of science, came to be seen as the primary definition of an age. In *L'évolution créatrice* the French philosopher Henri Bergson described humankind as *Homo faber* and claimed that future human evolution would express itself through evolving technical styles rather than being strictly biological, as with animals.[14] Reflecting the new technologies of electricity and chemicals, the Scottish prophet Patrick Geddes coined the phrase, "the second industrial revolution."[15] He thought that even this was only a passing stage, and he appropriated the idea of the "biotechnic age" from his German contemporary Raoul France. The long life of Geddes's protégé Lewis Mumford (1895–1990), author of *Technics and Civilization*, brings up to our own day his vision of a sequence of technology-based "ages," in which the eotechnic, paleotechnic, and neotechnic eras were succeeded by the "biotechnic."

As the place granted to biotechnics by philosophers of history suggests, biotechnology has been closely associated with hopes for industrial renewal throughout the century. As early as 1936 Julian Huxley predicted that "biotechnology will in the long run be more important than mechanical and chemical engineering."[16] That such hopes have not been better remembered is due, at least in part, to the confusion caused by two apparently very different strands of thought. Scholars have now shown that hopes that people could be improved by engineering their genetic inheritance dates to the beginning of this century.[17] This vision was itself so powerful, and its application by the Nazis so appalling, that its aura has often blinded historians to the other tradition—of hopes for an industry based on fermentation. Even though it has indeed often been pointed out the roots of biotechnology lie in brewing, this ancestry has tended to be treated more as an ancient curio than as a heritage with a modern significance.

From Zymotechnology to Biotechnology

During the later nineteenth century, an old, established industry and a visionary technology were linked through the movement for a progressive "zymotechnology"—an ever-broadening effort that began as a search for a better understanding of industrial fermentation, particularly beer. The brewing industry is important not merely for the comfort its product brings. In late nineteenth-century Germany it contributed as much to the gross national product as steel, and taxes on alcohol proved significant sources of revenue to government. Governments and entrepreneurs set up colleges for brewers. After the 1860s great institutes and remunerative consultancies were dedicated to the technology of brewing. The most famous was the private Carlsberg Institute, founded in 1875; the Carlsberg employed Emil Christian Hansen, who pioneered the pure yeast process for the reliable production of consistent beer. Less well known were private consultancies that advised the brewing industry. One of these, the Zymotechnic Institute, was established in Chicago by the German-born chemist John Ewald Siebel; it published a newsletter entitled (at least in 1884) the *Zymotechnic Magazine: Zeitschrift für Gährungsgewerbe and Food and Beverage Critic.* A similar name, though much more distinctive, adorned a Berlin institute known as the Institut für Gärungsgewerbe, which housed eighty scientists in a new building in 1909.

This movement's proudest hour came in World War I. The Berlin institute's director, Max Delbrück (uncle of the molecular biologist), had called yeast a wonder plant, and Germany grew it on an immense scale during the war to meet 60 percent of its animal feed needs.[18] Compounds of another fermentation product, lactic acid, made up for a lack of hydraulic fluid, glycerol.[19] On the Allied side the immigrant Russian chemist Chaim Weizmann (later the first president of Israel) used starch to eliminate Britain's shortage of a key explosives raw material by fermenting maize to acetone.[20] Clearly the industrial potential of fermentation was outgrowing its traditional home in brewing, and "zymo" gave way to "bio."

With shortages spreading and new grounds for hope needed, some dreamed of a new industrial base. Karl Ereky coined the word *Biotechnologie* in Hungary during 1917 to describe technology based on different biological raw materials. Ereky met the threat of wartime famine by fattening pigs intensively on an industrial scale: His fattening center handled 50,000 of the "biotechnologische Arbeitsmaschinen," as he termed the pigs, transforming an input of sugar beets into an output of lard and sausages.[21] In nearby Germany Delbrück's influential colleague Paul Lindner, editor of *Zeitschrift für Technische Biologie*, soon adopted the term.[22] Lindner was principally interested in using microbes as factories.

This catchword spread quickly after the war; it entered German dictionaries and was taken up abroad by business-hungry private consultancies as far afield as the United States and England. In Chicago, for example, the coming of Prohibition at the end of World War I ironically gave a boost to biological industries by creating opportunities for new fermentation products, in particular a new market for nonalcoholic drinks. Siebel, the Zymotechnic Institute's founder, was a lifelong opponent of the temperance movement. Several of his sons tried to diversify during Prohibition and carry on the consultancy, but one, Emil, broke away to establish a company called the "Bureau of Biotechnology," which specifically offered expertise in fermented nonalcoholic drinks.[23] In Leeds also a "Bureau of Biotechnology" that offered advice on applied microbiology was set up—this one by a consultant capitalizing on his expertise in the leather industry, which used microbiological methods to remove hair.[24]

The belief that the needs of an industrial society could be met by fermenting agricultural waste was an important ingredient of the so-called chemurgic movement. Devoted to finding novel uses for agricultural produce, the movement prospered during the agricultural depression of the late 1930s. The political pressure it exerted led to the formation of powerful organizations, such as the Regional Research Laboratories established by the USDA, and its economic arguments sustained research whose significance lasted long after.[25] Fermentation-based processes generated products of ever-growing utility. Penicillin was the most dramatic. Its biological activity was first observed in England, but it was produced industrially in the United States using a deep fermentation process originally developed in the federal Northern Regional Research Laboratory in Peoria, Illinois. Penicillin was more than just a drug among others. The enormous profits and the public expectations it engendered caused a radical shift in the standing of the pharmaceutical industry and suggested that fermentation had already inaugurated a utopia. Doctors used the phrase "miracle drug" without embarrassment, as hitherto-incurable conditions succumbed.[26] The historian of its wartime use, David Adams, has suggested that to the public penicillin represented the perfect health that went together with the car and the dream home of wartime American advertising.[27]

The 1950s brought steroids made by a fermentation technology. Cortisone in particular promised the same millennial ability to change medicine as penicillin had.[28] Even greater expectations were raised during the 1960s by a process that grew single-cell protein on paraffins. When the so-called protein gap threatened world hunger, producing food locally by growing it on waste seemed a solution. Major companies such as BP and ICI staked their future on it. The Soviet Union eventually produced more than 1.5 million metric tons a year on 60 percent of its liquid paraffin output.[29] In the late 1970s the U.S. government promoted the production of gasohol, gasoline with 10 percent alcohol added. At the end of the Carter administration, given the ban on exporting corn to the Soviet Union, fermenting agricultural surpluses to fuel seemed a neat and economical solution to the shortage of oil threatened by the Iran-Iraq war. As it industrialized, Brazil too embraced this new vision enthusiastically, seeing in fermented sugar cane an escape from dependence on imported oil. During the 1970s biotechnology therefore seemed to offer the chemical industry a new generation of technologies and a new image: clean, low-energy-consuming, and devoted to products highly relevant to human needs, particularly in the Third World.

This vision was often dimmed. Today it may appear that cortisone, single-cell protein, and gasohol are useful but not world-changing innovations. Later disappointments, however, do not prove earlier misguidedness. Moreover, that some promoted the vision cynically should not gainsay the idealism of many of its promoters. The journal entitled *Biotechnology and Bioengineering* since 1961 provided the focus for chemical engineers and microbiologists to discuss advances. Its editor, Elmer Gaden, worked tirelessly to promote this technology, which promised to counter the sadder uses of science and technology so notorious during the Vietnam War. With him through numerous committees worked J. W. M La Rivière of the Netherlands, now the head of ICSU, and Carl-Göran Hedén of Sweden, leader of the World Academy of Arts and Science. They created the sequence of meetings entitled Global Impacts of Applied Microbiology, first convened in 1962.[30] In Japan the Mitsubishi Chemical Company created its own Institute for Life Sciences in 1971, and a year later Bayer in Germany established its Biotechnikum Center.

In 1974 the German chemical industry association Dechema articulated the vision in an up-to-date report called simply *Biotechnologie.*[31] In an age of oil crises, the technology relied increasingly on the intersection of chemistry, chemical engineering, and microbiology, capitalizing on new abilities to ferment microorganisms in ways sensitive to the nature of organisms, the nature of raw materials, and the possibilities of engineering control. Products of this new synergy seemed to address the central needs of an age of scarcity: fuel, from fermented starch; food, for example single-cell protein grown on oil or glucose; medicines, antibiotics, steroids, and proteins; new extraction technologies, such as microbial mining of copper; fertilizers, in the form of rhizobia built into the roots of cereal to fix nitrogen. The report was addressed to the chemical industry, for biotechnology appeared to offer a non-hydrocarbon-based approach that would enable industry to survive. This belief inspired the foundation in 1978 of the European Federation of Biotechnology (EFB), with Dechema providing its secretariat, and initiatives by other European countries. As late as 1981 EFB adopted the following definition of biotechnology: "Biotechnology is the integrated use of biochemistry, microbiology, and engineering sciences in order to achieve the technological application of the capacities of micro-organisms, cultured tissue cells and the parts thereof."[32]

Biotechnology Old and New

In February 1974, a month after the Dechema report was published, *Fortune* magazine published an article on postwar microbe breeding that identified the bacterial geneticist Joshua Lederberg as one of six "men who are bringing you better microbes."[33] The article, appearing at the very moment of a new biotechnology, is a fascinating illustration of the excitement that in business circles preceded the exploitation of recombinant DNA techniques. The article foresaw three new directions in the field: the development of even better genetic techniques, a new emphasis on organisms important to industry rather than simply convenient for scientists, and new techniques for automated screening. The robotics of genetic engineering rather than the genetics received special emphasis: The Nobel Prize-winning physicist Don Glazer had recently developed his "Dumbwaiter," in which millions of colonies could be grown and studied simultaneously.

At the same time, the discipline of molecular biology was crystallizing on a variety of research fronts. Its first history, *Phage and the Origins of Molecular Biology*, was published in 1966. Although the principal schools in Britain and the United States actually developed in rather different ways, they shared a preoccupation with DNA and a reverence for its decoding by Francis Crick and James Watson in 1953. It has indeed been suggested that its role as a point of Anglo-American agreement gave the achievement a special historic significance.[34] DNA chemistry—and its reputation—grew quickly in the late 1960s. In December 1967, two months before the publication of Watson's *Double Helix,* Arthur Kornberg synthesized a viral gene. And in 1973 Stanley Cohen and Herbert Boyer managed to transfer a section of DNA from one *Escherichia coli* to another—a key event that would define the future of technological potential of molecular biology.

The new science of molecular biology, while apparently uncovering secrets of life, was not a product of traditional microbiology. Nor did its proponents have the close links to industry of the older discipline.[35] The achievements of the 1950s often seemed very abstract, and most of the large pharmaceutical companies were marginally, if at all, interested. When this new science came to appear useful after all, there was a profound social and economic rift between microbiologists and molecular biologists—and between practitioners in two industries. Established chemical companies may have incorporated recombinant DNA techniques within their armory of methods, but they did not promote them to the extent that enthusiastic small companies and entrepreneurial scientists did. Drug companies saw the techniques as simply one more tool to add alongside conventional screening for activity in natural organisms and chemical synthesis. Thus (ironically, in view of future developments and excitement across the Atlantic) the 1974 Dechema report hardly mentioned either genetics or molecular biology.

Molecular biologists too were first inclined to see new developments in their field not as opportunities, but rather as potential threats to the environment and to the autonomy of their discipline. The discoveries of the 1960s were raising widespread anxieties that were only heightened by an atmosphere of distrust—both of industry and of an older generation. Defoliation in Vietnam seemed to demonstrate to many how industry used science. Harvard graduate students discussing the social impact of genetic engineering from the late 1960s saw their science in danger of being used much as nuclear physics had been a generation earlier.[36]

During the 1960s it seemed that the greatest threat posed by the new genetics lay primarily not in its application to microbes — the threat to the environment — but its application to humans — the danger of a new eugenics. As the distinguished biologist Robert Sinsheimer admitted in a 1976 interview, "The problem is that when most people talk about genetic engineering they immediately leap — as I did myself a little earlier — to man, which is where all the really difficult moral problems arise."[37] As early as 1958 those members of society who wanted to hear were warned of the cultural challenges in prospect: That year the distinguished chemist Linus Pauling spoke of the threat of genetic disease on television, and E. L. Tatum, son of a distinguished pharmacologist and himself coauthor of the theory that integrated biochemistry with genetics, in his Nobel address alerted the world to a future governed by genetic technology.[38] Tatum described his vision again in 1963 at a meeting in Ohio where the very phrase "genetic engineering" may have been launched. His enthusiasm earned the riposte that American business would be more interested in exploiting genetic engineering for its profit than for its health-giving properties.[39]

A few months earlier Gregory Pincus — a former pupil of Jacques Loeb and a pioneer of that practical application of biology, the contraceptive pill — had encouraged London's Ciba Foundation to set up a meeting on the "future of mankind." It was attended both by an older generation of biologists, including Julian Huxley and J. B. S. Haldane, and by such younger representatives as Joshua Lederberg.[40] Lederberg rejected the appeal to eugenics and the manipulation of inheritance. He was medically trained and interested in practical treatments for inherited conditions. Rather than trying to modify germ cells, inheritance, and genotype, he concentrated upon the phenotype and cells that were not inherited. He expressed his faith in a new generation of therapeutics, which he called "euphenics," as distinct from the discredited "eugenics." Now that bacterial infections had been defeated by antibiotics, genetic diseases could be disarmed by genetic methods.

This vision alerted the pharmaceutical industry, and Lederberg himself so impressed Carl Djerassi of the Syntex Corporation, a pioneer of steroid manufacture, that he gave Lederberg a research laboratory to explore its implementation. Lederberg later (in the early 1970s) persuaded the then-new Cetus Corporation to use the new technique to engineer other useful bacteria. At the British subsidiary of Searle, another American pharmaceutical firm, two biochemists, Brian Richards and Norman Carey, were so impressed by the potential of the new molecular biology in 1967 that they wrote a notarized document on the potential uses of fused DNA from different organisms were it to prove possible.[41] They later helped found key British biotechnology companies. They foresaw how gene technology might implement early hopes for biotechnology, but in the early 1970s their optimism was still unusual.

Asilomar and Later

Scientists are not generally given to expressions of dreams and evocations of the untried. Robert Teitelman has described how in the 1980s scientists found that grand promises attracted money, and how in turn many were attracted by the enormous rewards (both personal and industrial) that promised to follow from applying molecular biology.[42] In the early 1970s, however, personal financial ambition on that scale lay in the future; it cannot explain why many expressed such hopes for applied molecular biology. Academic practitioners of a rather esoteric science did not begin by believing that they were about to move personally to the more remunerative but also more secretive world of business. Instead, scrupulous scientists were driven to express their claims urgently and vociferously by the threat that the new techniques would be banned both in the research laboratory and in industry. In defending scientific work by countering talk of the threats of the new biology, they coopted those promises long proffered by biotechnology and eventually the world itself.

After the initial discovery of genetic recombination, the small and closely interconnected molecular biology community decided to impose a moratorium on such work rather than rush to new experiments. That voluntary moratorium, initiated by a letter to *Science* and to *Nature* in 1974, lasted until NIH promulgated new regulations in June 1976.[43] It has been widely pointed out that this moratorium was defensive: to assure the public that the molecular biologists were not mindlessly creating Frankensteinian monsters. What is less often noted is that several of the scientists saw the moratorium as only one prong of the defensive campaign. The other prong was to show the potential benefits that would flow when it was lifted.

For Lederberg the attitudes behind the moratorium were much too negative. At a conference set up at Asilomar, California in February 1975 to create a professional consensus on safety and prospects, he circulated a paper expressing his own then-unpopular viewpoint. He suggested that the new techniques offered "an early chance for a technology of untold importance for diagnostic and therapeutic medicine: the ready production of an unlimited variety of human proteins. Analogous applications may be foreseen in fermentation processes for cheaply manufacturing essential nutrients, and in the improvement of microbes for the production of antibiotics and of special industrial chemicals."[44] Two features of this enumeration of speculative benefits are striking, especially in view of later rhetoric, which appealed to the private profit promised by cutting-edge technology. Lederberg intended not to raise money, but to convince people not to overregulate science in order to avoid distant risks. The list was also very conventional in the light of fifty years of biotechnology. It was therefore as if recombinant DNA techniques were enabling the fulfillment of long-held hopes. This focus on benefits rather than risks was unusual. It led to accusations that Lederberg was ignoring the social dimension of molecular biology, so that he felt driven to write to the *New York Times* to rescue his reputation as a socially concerned scientist.[45] Yet his list would become a model statement of the potential benefits of biotechnology and in various forms would inform presentations for funds and public support.

Not just Lederberg, but also microbiologists, the traditional progenitors of biotechnology, were isolated at Asilomar, insofar as they were present at all. Those traditional guardians of pathogenic organisms were completely sidelined—and they resented it.[46] The most remarkable expressions of this anger were to be found in England. Ernst Chain, one of the pioneers of the work on penicillin, on retiring from his chair of biochemistry at London's Imperial College threatened to sue the college when he found that he was to be succeeded by a molecular biologist.[47] It took all the power of England's elite to dissuade him. When Paul Berg, the first signatory of the cautionary letters to *Nature* and *Science*, led a televised debate on the research moratorium in London's Royal Institution, he was vehemently criticized by John Pirt, a distinguished microbiologist and a pupil of Chain, who asserted that skilled microbiologists were quite able to handle pathogens.

Perhaps one reason for their irritation was that even then Berg himself was beginning to coopt the claims of biotechnology for his own discipline in defending the latter against punitive controls. He asked rhetorically: "Why can't these simple organisms become the factory for producing some of society's most needed supplies—antibiotics—hormones—and even a source of food. And for those who enjoy real speculation there are of course the dramatic possibilities for introducing new genes into human cells and thereby trying to cure certain genetic diseases."[48] Berg did two things in this telecast. First, he listed the benefits associated with biotechnology, essentially the same ones the *Fortune* article had listed seven months earlier. Second, he shifted his tone from the microbiologists' optimism. The stakes had been raised. The reason was clear as he went on to balance distant benefits with equally uncertain risks, such as developing antibiotic resistant bacteria, or converting harmless into toxic organisms. Thus as early as 1974 a list of speculative gains was being set against the list of speculative risks, though then the issue was cast as dangers counterbalancing benefits.

The debate over academic research was partially resolved by identifying some processes as very low risk; these were controlled by the NIH regulations enacted in mid-1976. Even then there were debates in many cities and on many campuses about the wisdom of establishing genetic research laboratories. Moreover, the NIH legislation did not address industrial research or production. By mid-1977 there were sixteen bills before Congress concerned with controlling genetic engineering research and production.

Again supporters of biotechnology argued for its potential benefits. An interview with Irving Johnson, research director of Eli Lilly, was printed in the company's newsletter *Lilly News* and submitted in full to a congressional hearing; the interview was explicitly targeted at the anxieties and "misunderstandings" evident in Congress. The debate was interesting not for its originality but rather for its formulaic quality. In defending his industry's research, Johnson argued for projected benefits—still theoretical—that would range from combination tomato and potato plants to cures for genetic diseases.[49] Although the threat of immediate congressional action faded, the Office of Technology Assessment was asked to provide the background report for Congress that Harsanyi, then a professor of genetics at Cornell, headed. That report, submitted in 1981, carefully set speculative benefits against equally speculative threats.[50] By then of course the mood in America had changed completely, yet the arguments maintained their particular style in popular books and articles with titles like *Life for Sale*. The chapter titles of this well-known 1981 survey give the flavor: "Turning DNA into Gold," "Biobusiness Takes Off," "A New Era of Miracle Drugs," "The Next Industrial Revolution," "Feeding the World," "The Age of Human Genetics," "Is it Safe? Revisiting the Great Debate," and "Who Owns Life, and Other Easy Questions."[51] Biotechnology had been presented in this light for so long that its wonder-boy status was then taken for granted.

Biotechnology on Wall Street

The matching of costs and benefits in the evaluation of biotechnology was often expressed by analogy with one of two contrasting models: information technology and nuclear power. Information technology was a positive model. The United States, while losing out in traditional areas such as the steel industry, was seen as achieving world dominance in information, with shareholders and venture capitalists profiting enormously. In Europe, too, that analogy seemed to fit biotechnology. The European Commission authorized a report on industry in the future that postulated that society would be based on information technology in the short term, then on biotechnology.[52] The other frequently used model was nuclear power—another high technology that had raised great hopes at its inception, but by the late 1970s appeared to offer more threats than benefits. In Germany many of the protesters against biotechnology in the 1980s had cut their teeth on the earlier antinuclear movement.[53]

Wall Street decided that biotechnology more resembled information technology. At least one crucial moment in biotechnology's transition from being the subject of anxious comment to wonder-boy status can be pinned down exactly. In June 1979 Nelson Schneider, an analyst for E. F. Hutton, heard the exciting news that a small new company called Genentech had reported production of human insulin. To find out more about a new biology that had allowed a new firm to outplay such an established corporation as Eli Lilly, he sought the advice of Cetus's Pete Farley, who recommended that he attend a meeting at London's Royal Society. The meeting dealt with the implications of recombinant DNA, but located the work firmly within the microbiological conception of biotechnology.[54] Schneider returned, reinterpreting the London message: here was a completely new technology, he reported in a letter to investors. Yet he maintained the breadth of perception long associated with biotechnology. He recognized the publicity regularly evoked by pharmaceuticals: "In our judgment, however, although the pharmaceutical industry has done some of the most advanced pioneering work on DNA, the most exciting investment potential could lie in energy production and agriculture."[55] Schneider's vision of a new generation of technologies reflected Delbrück's a century before. "The cost savings possible from manufacturing fermented products through DNA technology can be staggering because the self-generation of the product's cell organisms may be as close as science has approached to a perpetual motion machine."[56]

Schneider organized a meeting in Washington on 17 September 1979 to present a few papers by Harsanyi, academic scientists, and industry people. Instead of the few dozen participants expected, hundreds showed up. Schneider had been outstandingly successful in marketing an investment concept. The new revelation would go under the name *biotechnology*, and great would be its profits. Schneider later testified to Congress that in this area lay the roots of a new IBM. His company, E. F. Hutton, copyrighted the word *biotechnology* as the title of a newsletter for investors, one dealing with applied genetics rather than with biotechnology in the established, broader meaning. Hutton, of course, had a strong financial incentive for separating the new investment opportunity from earlier, more financially mature technologies. So did investors who planned to exploit hopes for both speculative profits and great dividends. Their financial interests were reflected in the emergence of numerous companies (many of them funded by companies such as Hutton) whose expertise lay specifically in molecular biology rather than in the older trades of fermentation engineering or microbiology.

The distinction between the old and new biotechnology began with the distinction between the microbiologists and the molecular biologists; it was strengthened by the novelty of the new DNA-modifying techniques and by the wish to identify biotechnology as a new technology. In practice, the new biotechnology would draw upon more than the visions held by an older generation. Additional established skills were required, too, and they were often brought in from other disciplines. In any case, pharmaceutical and agrochemical applications were increasingly emphasized, so that today the distinction between old and new biotechnologies may already be an irrelevance: the continuities are now much more evident than they were a decade ago.

While the new biological and biochemical techniques have transformed the range of what is now possible scientifically and practically, I have argued nonetheless that the genetics-based biotechnology that emerged in the 1970s drew its inspiration from the established rhetorical tradition of biotechnology—and drew the urgency of its own rhetoric from the perceived need to defend infant research from control. Not that the participants engaged in conspiracy, or even intended to change the terms of the debate to the degree that occurred. That change owes much to the stimulus provided by market analyses promising turnovers in the billions of dollars within a few years, and to the capital gains tax laws that encouraged American venture capitalism. Nonetheless, for stock-market analysts like Schneider and for journalists, performances such as those of Berg, Lederberg, and Harsanyi defined biotechnology as an industry with an overwhelming potential and provided the urgent sense that this potential could be realized now.

Notes

1. Victor Hugo, *Les Misérables* (Paris, 1862), trans. Norman Denny (London: Penguin, 1982), p. 19.

2. John Durant, ed., *Biotechnology in Public: A Review of Recent Research* (London: Science Museum, for European Federation of Biotechnology, 1992).

3. John Elkington, *The Gene Factory: Inside the Biotechnology Business* (London: Century Publishing, 1985), pp. 44–45.

4. The respected British magazine *Chemistry in Britain* suggested of biotechnology, "Perhaps the whole concept should be abandoned, the relevant techniques simply regarded as part of chemistry or biochemistry." "Life Study," *Chemistry in Britain* 30 (1994), 20.

5. For the fears see Sheldon Krimsky, *Biotechnics and Society: The Rise of Industrial Genetics* (New York: Praeger, 1991). See also Krimsky's article in this volume.

6. There is a rich literature on American utopianism and technology. See, e.g., Joseph J. Corn, ed., *Imagining Tomorrow: History, Technology, and the American Future* (Cambridge, Mass.: MIT Press, 1986); and Howard P. Segal, *Technological Utopianism in American Culture* (Chicago: University of Chicago Press, 1985).

7. Charles Rosenberg, "Introduction: Framing Disease: Illness, Society, and History," in *Framing Disease: Studies in Cultural History*, ed. Rosenberg and Janet Golden (New Brunswick, N.J.: Rutgers University Press, 1992), pp. xiii–xxvi.

8. The way the word itself acts as a boundary object between "bio" and "technology" is explored in Robert Bud, "Biotechnology in the Twentieth Century," *Social Studies of Science* 21 (1991), 415–57.

9. See, e.g., *Der grosse Brockhaus*, 15th ed. (1929); and *Meyers Lexikon*, 7th ed. (1925).

10. Zsolt Harsanyi, "Biotechnology and the Environment: An Overview," in *Biotechnology and the Environment: Risk and Regulation*, ed. Albert H. Teich, Morris A. Levin, and Jill H. Pace (Washington D.C.: AAAS, 1985), pp. 15–27.

11. U.S. Congress, House of Representatives, Committee on Energy and Commerce, Subcommittee on Oversight and Investigations, *Biotechnology Regulation*, 98th Cong., 2nd sess., 11 Dec. 1984, p. 1.

12. For a critical report on the concept of a new industrial society see Michael Adas, *Machines as the Measure of Man: Science, Technology, and Ideologies of Western Dominance* (Ithaca, N.Y.: Cornell University Press, 1989).

13. Clarence King, "The Education of the Future," *The Forum* 13 (1892), 20–33, 27. I am grateful to Philip Pauly, who posted this quotation on the Htech-L bulletin board.

14. I have explored Bergson's significance in greater detail in Robert Bud, *The Uses of Life: A History of Biotechnology* (Cambridge: Cambridge University Press, 1993), pp. 54–55. See also Jean-Pierre Séris, "Bergson et la technique," in *Bergson: Naissance d'une philosophie* (Actes du Colloque de Clermond-Ferrand, 17 et 18 novembre 1987) (Paris: Presses Universitaires de France, 1990), pp. 121–38.

15. See Patrick Geddes, *Cities in Evolution: An Introduction to the Town Planning Movement and the Study of Civics* (London: Williams and Norgate, 1915), p. 59.

16. Julian Huxley, "Chairman's Introductory Address," in Lancelot Hogben, *The Retreat from Reason: Conway Memorial Lecture, Delivered at Conway Hall, May 20, 1936* (London: Watts, 1936), p. vii. Huxley's meaning is explored in Bud, *Uses of Life* (note 14), pp. 71–77.

17. Daniel J. Kevles, *In the Name of Eugenics: Genetics and the Uses of Human Heredity* (New York: Viking, 1983); and Peter Weingart, Jürgen Kroll, and Kurt Bayertz, *Rasse, Blut, und Gene: Geschichte der Eugenik und Rassenhygiene in Deutschland* (Frankfurt: Suhrkamp, 1988).

18. Max Delbrück, "Hefe ein Edelpilz," *Wochenschrift für Brauerei* 27 (1910), 373–76; and R. Braude, "Dried Yeast as Fodder for Livestock," *Journal of the Institute of Brewing* 48 (1942), 206–12.

19. H. Benninga, *A History of Lactic Acid Making: A Chapter in the History of Biotechnology* (Dordrecht: Kluwer, 1990).

20. Robert Bud, "The Zymotechnic Roots of Biotechnology," *British Journal for the History of Science* 25 (1992), 127–44.

21. For the first expression of his philosophy see Karl Ereky, "Grossbetriebsmäßige Entwicklung der Schweinemast in Ungarn," *Mitteilungen der Deutschen Landwirtschaftlichen Gesellschaft* 34 (1917), 541–50. See also Ereky, *Biotechnologie der Fleisch-, Fett-, und Milcherzeugung im landwirtschaftlichen Großbetriebe* (Berlin: Paul Parey, 1919).

22. Paul Lindner, "Allgemeines aus dem Bereich der Biotechnologie," *Zeitschrift für Technische Biologie* 8 (1920), 54–56.

23. E. A. Siebel and Company and Siebel Laboratories, Inc., *Achievement: Yesterday, Today, Tomorrow* (Chicago, n.d.).

24. See Bernard Dixon, "Putting the 'Bio' in Biotech," *New Scientist* (31 Jan. 1985), 38.

25. The classic article is Carroll Pursell, "The Farm Chemurgic Council and the United States Department of Agriculture, 1935–1939," *Isis* 60 (1969), 307–17. See also Peter Neushul, "Science, Government and the Mass Production of Penicillin," *Journal of the History of Medicine and Allied Sciences* 48 (1993), 371–95.

26. David Tyrell, private communication, 8 Feb. 1994.

27. David P. Adams, "The Penicillin Mystique and the Popular Press (1935–1950)," *Pharmacy in History* 26 (1984), 134–42.

28. For the understudied history of cortisone see Norbert Klinkenberg, *Cortison: Die Geschichte des Cortisons und der Kortikosteroidtherapie: Ein Beitrag zur Forschungs- und Therapiegeschichte heutiger Medezin* (Cologne: Pahl-Rugenstein, 1987). The fermentation route is described in Durey H. Peterson, "Autobiography," *Steroids* 45 (1985), 1–17. On the general history of fermentation technologies see also David Perlman, "Fermentation Industries . . . Quo Vadis," *Chemtech* 7 (1977), 434–43; and David Perlman, "Stimulation of Innovation in the Fermentation Industries, 1910–1980," *Process Biochemistry* 13 (May 1978), 3–5.

29. Anthony Rimmington with Rod Greenshields, *Technology and Transition: A Survey of Biotechnology in Russia, Ukraine, and the Baltic States* (London: Pinter, 1992);

and David H. Sharp, *Bio-Protein Manufacture: A Critical Assessment* (Chichester: Ellis Horwood, 1989).

30. Carl-Göran Hedén, "The GIAMS: A Contribution to Technology Transfer" in *From Recent Advances in Biotechnology and Applied Biology*, ed. S. T. Chang, K. Y. Cian, and N. Y. S. Woo (Hong Kong: Chinese University Press, 1988), pp. 63–74.

31. *Biotechnologie: Eine Studie über Forschung und Entwicklung—Möglichkeiten, Aufgaben und Schwerpunkte der Förderung* (Frankfurt: Dechema, 1974). For the classic account of the origins of this report see the analysis by the secretary of the working group, Klaus Buchholz, "Die Gezielte Förderung und Entwicklung der Biotechnologie," in *Geplante Forschung*, ed. Wolfgang van den Daele, Wolfgang Krohn, and Peter Weingart (Frankfurt: Suhrkamp, 1979), pp. 64–116.

32. *EFB Newsletter* (European Federation of Biotechnology), Dec. 1981, No. 4, p. 2.

33. Gene Bylinsky, "Industry Is Finding More Jobs for Microbes" *Fortune*, Feb. 1974, 96–102.

34. I am indebted to Soraya de Chadaravian for allowing me to cite her concept of the strategic place of the double helix in the shared history of molecular biology, from "Building Molecular Biology in Cambridge (1947–1962)," a paper presented to the History of Twentieth-Century Medicine Group at the Wellcome Institute in London, 9 Nov. 1993. See also John Cairns, Gunther S. Stent, and James Watson, eds., *Phage and the Origins of Molecular Biology* (Cold Spring Harbor, N.Y.: Cold Spring Harbor Laboratory, 1966.).

35. See Lily E. Kay's article in this volume; and Kay, *The Molecular Vision of Life: Caltech, the Rockefeller Foundation, and the Rise of the New Biology* (Oxford: Oxford University Press, 1993).

36. Richard Roblin, interview by Charles Weiner, 21 April 1975, pp. 9–14, MC100, MIT Institute Archives and Special Collections, MIT Libraries, Cambridge, Mass.

37. Robert Sinsheimer, interview by Charles Weiner, 26 Dec. 1975, p. 42, MC100, MIT Archives.

38. For Pauling's telecast, "The Next Hundred Years" (13 Dec. 1958), see Kay, *Molecular Vision of Life* (note 35), 274–75.

39. Rollin D. Hotchkiss, "Portents for a Genetic Engineering," *Journal of Heredity* 56 (1965), 197–200.

40. See Gordon Wolstenholme, ed., *Man and His Future: A Ciba Volume* (London: J. A. Churchill, 1963). This meeting is discussed at greater length in this volume by Lily Kay.

41. B. M. Richards and N. H. Carey, "Insertion of Beneficial Genetic Information," Searle Research Laboratories, 16 Jan. 1967. I am grateful to Dr. Richards for the chance to see this document.

42. Robert Teitelman, *Gene Dreams: Wall Street, Academia, and the Rise of Biotechnology* (New York: Basic Books, 1989).

43. Among the many accounts of the Berg letter and the subsequent Asilomar Conference see Sheldon Krimsky, *Genetic Alchemy: The Social History of the Recombinant DNA Controversy* (Cambridge, Mass.: MIT Press, 1982); and June Goodfield, *Playing God: Genetic Engineering and the Manipulation of Life* (London: Hutchinson, 1977).

44. Joshua Lederberg, "DNA Research: Uncertain Peril and Certain Promise," *Prism* (AMA Policy Journal), 15 June 1975.

45. Joshua Lederberg, "A Geneticist on Safeguards," *New York Times*, 11 March 1975.

46. E. S. Anderson (of the British Public Health Laboratory Service), interview by Charles Weiner, pp. 43–56, MC100, MIT Archives (note 36).

47. Ronald Clark, *The Life of Ernst Chain: Penicillin and Beyond* (London: Weidenfeld, 1985).

48. BBC, "Certain Types of Genetic Research Should Be Suspended," BBC 2 broadcast, 16 Sept. 1974.

49. U.S. House Committee on Science and Technology, Subcommittee on Science Research and Technology, *Science Policy Implications of DNA Recombinant Molecule Research*, 95th Cong., 1st sess., 1977, pp. 474–79.

50. U.S. Congress, Office of Technology Assessment, *Impacts of Applied Genetics, Micro-Organisms, Plants, and Animals* (OTA-HR-132) (Washington, D.C.: U.S. Government Printing Office, 1981).

51. Sharon McAuliffe and Kathleen McAuliffe, *Life for Sale* (New York: Coward, McCann, and Geoghegan, 1981).

52. FAST Group, *Eurofutures: The Challenge of Innovation* (London: Butterworth, 1987).

53. Wolfgang Rüdig, *Antinuclear Movements: A World Survey of Opposition to Nuclear Energy* (London: Longman, 1990).

54. S. Brenner, B. S. Hartley, and P. J. Rodgers, eds., "New Horizons in Industrial Microbiology," *Philosophical Transactions of the Royal Society* B 290 (1980), 277–430.

55. Nelson Scheider, "DNA—The Genetic Revolution," *E. F. Hutton Research*, 1 Aug. 1979.

56. Schneider, "DNA—The Genetic Revolution" (note 55), p. 3.

Problematizing Basic Research in Molecular Biology

Lily E. Kay

In the 1970s the term *biotechnology* burst into the global arena seemingly sui generis, owing its novelty and allure to its origins in molecular biology and the nascent techniques of recombinant DNA. Defined as "the application of scientific and engineering principles to the processing of materials by biological agents to provide goods and services," the term (actually coined around 1917) is now embedded in a politics of history and meaning: either touted for its promise as a revolutionary technology or, in response to alarm over risks, downplayed as age-old domestication of nature. Indeed, this politicization of the term *biotechnology* caused key officials at the National Science Foundation (NSF) and the Food and Drug Administration (FDA) to feel that the word has become a "significant millstone around the neck of both the industry and the government," and to urge its removal from the regulatory vocabulary.[1]

Clearly, the new linkages between the engineering techniques of molecular genetics and the technologies of domesticating animate nature have tended to simplify and amplify the discontinuities and novelties in the relations between the molecular life sciences and the technologies they have spawned. As Robert Bud has pointed out in *The Uses of Life* (and again in this volume), the entangled roots of modern research and commerce, of biotechnology and molecular life sciences (e.g., fermentation, food and drug processing, plant and animal breeding, genetics, microbiology, biochemistry), reach back to the early twentieth century.[2] Similarly, elsewhere in this volume, Angela Creager shows how the patronage of molecular biology during and after World War II—and the concurrent changes in the relationships between the federal government, university research laboratories, and pharmaceutical companies—set the stage for the reception of the scientific innovations associated with genetic engineering. She makes the point that this kind of consolidation of cognitive, political, and commercial power predates by several decades the university-industrial complex based on recombinant DNA. While these studies may not prove a general trend, they do question how sharp the boundaries are between the pre- and post-recombinant DNA eras and between pure and applied research. They point to the commodification and market value of early molecular biology.

My article too questions whether sharp cleavages exist between the pre- and post-recombinant DNA eras. By viewing the rise of molecular biology as a cultural project and broadening the meaning of market value, it challenges a one-dimensional picture of basic research as distinct from its application and problematizes the perceived linear movement from the pure to the applied. By focusing on several important moments in the history of molecular biology from the 1930s through the 1960s, I argue that, from its very inception around 1930, the molecular biology program was defined and conceptualized in terms of technological capabilities and social possibilities; hence its enormous appeal. The ends and means of biological engineering were inscribed into the molecular biology program. These inscriptions—these cognitive commitments and social expectations—were captured in the changing approaches to heredity, cell, soma, and psyche. As inscriptions they were manifested in altered modes of representing and speaking about life—that is, as discursivities; and in altered modes of doing and intervening—that is, as instrumentalities.

This usage of the terms *representing* and *intervening* is not novel. They are the foci of analysis of a study by Ian Hacking in which he proposes that we view reality as intervention and see science as doing, rather than merely knowing. Reality, in Hacking's framework, is what we can use to intervene in the world to affect something else, or what the world can use to affect us—a mutual inscription process. He traces the interlocking of representing and intervening to the Baconian program, to the birth of an autonomous experimental tradition, whose primary aim was to manipulate and control nature for the utility of man and to collapse the dichotomy of the natural and the artificial.[3]

Several scholars, among them Hacking, Paul Forman, and Evelyn Fox Keller, have examined the implications of this thesis for the project of pure science, and for the long-cherished ontological cleavage between the pure and applied. They conclude that the interventionist drive in basic research has shaped the types of applications that emerged.[4] This line of thinking—that interventionist goals shape representational strategies—informs my own analysis of molecular biology as an interventionist program, a project whose applications were already inscribed in its design. Thinking along these dimensions—inscriptions, discursivities, instrumentalities—enables us to view the molecular biology project as both a scientific and a cultural enterprise, broadening and deepening the notion of commodification and market value. Viewed from this vantage point, the continuities between the 1930s and 1970s become as remarkable as the cleavages.

Life as Technology: Molecular Biology, 1930s–1940s

The role of the Rockefeller Foundation in the rise of molecular biology has been amply documented; several studies have offered a variety of perspectives on that social-scientific process.[5] It is worth a reminder that molecular biology was a "private science" for the first quarter century of its founding. Throughout the 1930s and 1940s it was supported by the private sector and flourished principally in elite (private) academic institutions. Big business did not directly support molecular biology, but did sponsor it indirectly through the Rockefeller Foundation, whose trustees and officers shared many of the corporate goals of industrial capitalism: its economic premises, political commitments, and social visions. To understand the rise of molecular biology within this political economy, the Rockefeller Foundation's program must be situated within the Foundation's agenda "Science of Man." The motivation behind the enormous investment in the new agenda was to develop the human sciences as a comprehensive explanatory and applied framework of social control grounded in the natural, medical, and social sciences. Indeed, all three areas were part of the rehabilitation project to promote empirical, objectivist, experimental, and atomistic methods modeled after the physical sciences.[6]

Raymond B. Fosdick, chief architect of the new agenda and president of the Rockefeller Foundation (1936–1948), argued that there was an urgent need for "the same kind of fearless engineering in the social field that in the realm of physical science has pushed out so widely the boundaries of human understanding." Like most of his peers at the foundation and in academe, Fosdick appealed throughout the 1920s to rational social control based on biological and eugenic principles: "Can the conscious effort of men in any way steer this biological evolution? Is it possible to adjust the size of population to fit the world's resources so that those who inhabit the earth can do so in seemliness and dignity? Can we shift the emphasis from quantity of human life to quality of human life? Can the science of eugenics reshape a process that is tumbling with such gigantic forces?"[7]

As a means of social control, the human sciences were to supply instrumental rationality to the managerial sector. From its inception, the rationale for the technology-based program of molecular biology, and for its residual eugenic goals, was its future social returns: not so much immediate commercial applications (though such activities were applauded), but a long-term promise of generating *social technologies*. As an internal foundation report put it,

> The challenge of this situation is obvious. Can man gain intelligent control of his own power? Can we develop so sound and extensive a genetics that we can hope to breed, in the future, superior men? Can we obtain knowledge of the physiology and psychobiology of sex so that man can bring this pervasive, highly important, and

dangerous aspect of life under rational control? Can we unravel the tangled problem of the endocrine glands, and develop, before it is too late, a therapy for the whole hideous range of mental and physical disorders which result from glandular disturbances? Can we solve the mysteries of the various vitamins so that we can nurture a race sufficiently healthy and resistant? Can we release psychology from its present confusion and ineffectiveness and shape it into a tool which every man can use every day? Can man acquire enough knowledge of his own vital processes so that we can hope to rationalize human behavior? Can we, in short, create a new science of man?[8]

These science-driven visions should not be viewed as Machiavellian forms of social control. It is crucial to underscore that the Rockefeller Foundation's support of geneticists, biochemists, biophysicists, immunologists, and microbiologists within the "Science of Man" agenda did not constitute academic subversion and cooptation. The complex set of relations of scientists to patrons and of intellectual programs to the social agenda is productively explained within an analytic framework of cultural hegemony, through the explicit and tacit constitutive processes of consensus formation. Within that framework, "power" includes intellectual, cultural, political, and economic power, and mental life is not a mere shadow of material life. From this perspective the maintenance of hegemony does not require active commitment by an academic constituency (or by subordinates) to legitimate elite rule. Rather, subordinates and rulers, academics and business elites reinforce each other in a circular manner to form a "hegemonic bloc" sustained by formal and informal systems of incentives and power sharing, particularly through half-conscious modes of complicity. Hence this view regards hegemony not as a form of subtle coercion or top-down social control, but rather as an interactive process between social groups vying for power.[9]

Within this framework of cultural hegemony it does not matter that many of the scientists funded through the Rockefeller Foundation's molecular biology program were unaware of their social function as laid out in the "Science of Man" agenda. The leaders of American life science — many of them acting as scientific advisers to the foundation — did understand the larger picture; but even they did not always share all the goals of their patrons. The rise of the new biology was a process of consensus building among interdependent though not identical professional constituencies with common as well as separate goals. In search of patronage, however, most leaders of pure science did argue for their service role to the political economy. Thus consensus can form without willed complicity. Whether practitioners of molecular biology valued their work as contributions to a broad social agenda, or through perceived intellectual autonomy they deployed the appropriate rhetoric to fund their basic research, is secondary to the process of consensus formation (though at the highest levels of management there was substantial agreement). What is of primary importance is that through their authority scientists did empower the Rockefeller program. By offering expertise they supplied an instrumental rationality that not only legitimated their own enterprise but also validated the cultural objectives of their patrons.[10]

Thus molecular biology qua human science became central to twentieth-century notions of what Michel Foucault termed "bio-power," or power over life: control of bodies, and control of populations. The processes and products of that sociotechnical system that we call "science" and "technology" have worked because they have been embedded in cultural practices. They have been stabilized and naturalized in our technologies for producing truths, truths circulated through an economy of discourse.[11]

With the rise of molecular biology we observe an emergence and eventual stabilization of certain sociotechnical truths; we witness an economy of discourse in the way that certain phenomena in the 1930s came to be systematically configured together. For example, the term *cooperation* represented both a managerial and a cognitive prescription; social control was aimed at rationalizing and controlling individual and group behavior; behavior (as representation of personality and intelligence) was seen as shaped by biological processes; biological processes were increasingly conflated with notions of genetic determinism; genes were taken to be physical entities whose materiality was constituted by protoplasmic endowment; and protein action was thought best studied on the molecular level, through molecular technologies that are at once representations and interventions. These discursive configurations provide an excellent example of such a socioscientific stabilization, of the economy of discourse. Within this framework of bio-power, discursivities and instrumentalities, representations and interventions, are viewed as two sides of the same coin.[12]

True, in the 1940s the manipulations and representations of hereditary mechanisms were carried on principally at the submicroscopic level of macromolecules and were guided by minimalist experimental systems such as fungi, bacteria, viruses, and phages—microorganisms separated by great phylogenetic distances from humans and by an even greater distance from modes of social control. But those who deployed these molecular probes and the streamlined model systems, and convinced the Rockefeller officers of their fruitfulness—Max Delbrück (phage research), George Beadle and Edward L. Tatum (*Neurospora* studies), Linus Pauling (protein structure), and Joshua Lederberg (*E. coli* research), to name several key actors on the American scene—readily extrapolated from microorganisms and macromolecules to humans. Scientists and patrons came to share a molecular vision of life. As such they became coproducers of a discourse that represented organisms as the genetically directed activity of molecules and viewed the study of microorganisms and proteins as the surest path to understanding and controlling human physiology. Though not an applied science, molecular biology in the 1930s and 1940s was mission-oriented.

Even within these long-range visions and social justifications, commercial applications were not far from view. And when commercial opportunities arose in the 1940s they were vigorously pursued; the market value of molecular biology and the commodification of molecules and organisms were surely recognized. Two examples come readily to mind: the contingencies surrounding Beadle's *Neurospora* work at Stanford and the opportunism of Pauling's research on antibodies at California Institute of Technology.

Beadle became widely known in the community of life scientists in the 1940s for his outstanding contributions to biology, linking formalistic concepts of classical genetics with biochemical explanations. By using a relatively simple fungus—the bread mold *Neurospora crassa*—to probe the gene, Beadle, a geneticist, and Tatum, a biochemist, managed to disentangle some of the long-standing circularities in the relation between genes and enzymes. These studies led to the celebrated, "one gene—one enzyme" hypothesis and brought American genetics and biochemistry closer together.[13]

What made Beadle's program even more remarkable was its connection to the war effort. His *Neurospora* research was launched at the end of 1940, just at the height of the "preparedness period," and reached its zenith in 1943, just when most fundamental researches were being cut back. Although Beadle's primary commitment was to fundamental research, it was mainly the practical applications that gave his program its lead in the competition for priority and resources: industrial liaisons, government contracts, and military deferments. Undoubtedly because Tatum had studied at the University of Wisconsin (and his father had worked there, with close ties to pharmaceutical firms), Beadle linked his research to the food and drug industries. The combination of *Neurospora* genetics with methods of nutrition, especially the *Neurospora* bioassay, proved attractive procedures for manufacturing vitamins and amino acids and caught the notice of such institutions as Merck, Sharp and Dohme, the Fruit Product Laboratory, and the Western Regional Department of Agriculture. When Beadle returned to Caltech in 1947 as chairman of the biology division, he had proved himself (much as Edwin Cohn did in Creager's case study in the Harvard war project) a savvy manager of university, industry, and government interests—an equal to Pauling.[14]

Pauling too, in spite of a deep commitment to fundamental research, knew how to harness the commercial potential of his work. His immunochemistry research, one of the pillars of molecular biology within the protein paradigm, promised to effect a commercial revolution in medicine and agriculture. Pauling's instructive theory of antibody formation, which postulated a mechanism by which an antigen determined the complementary configuration of the polypeptide chains comprising the antibody, dominated the thinking in immunology into the mid-1950s. Its appeal lay not only in its apparent coherence and explanatory power (including explaining the action of the protein gene), but in its revolutionary implication for physiology and medicine: Pauling's scheme suggested that any antibody derived from serum or globulins could be manufactured in vitro.[15]

The new technology must have seemed limited only by the imagination. Not only would it enable humans and animals to ward off diseases simply with an infusion of artificial antibodies, it could enable scientists to alter the immune system and even an organism's genetic endowment. If patented, the procedure for making artificial antibodies would mean a windfall of profits to Caltech and pharmaceutical companies and render Pauling's immunochemistry one of the most scientifically successful and commercially lucrative projects in history. In the summer of 1941 Pauling quietly applied for a patent on artificial manufacture of antibodies. That promise, though it failed, paid. Throughout the 1940s, despite negative results, the immense potential of the process attracted funding from commercial interests and the Rockefeller Foundation.[16]

These admittedly limited forays into commerce proved important for the organization of postwar life science. They broadened the resource base of molecular biology. As the Rockefeller support of U.S. science declined, government contracts, private foundations, pharmaceutical companies, and agricultural industries began sponsoring various aspects of molecular biology. These ventures into the marketplace (like those documented by Bud and Creager) were not isolated instances but manifested the stable commercial infrastructure for the molecular life science that dated back to the 1920s. More significantly, these instances were not merely precursors of commodification, they extended and deepened preexisting inscriptions of molecular biology as a technology of life. These commercial ventures enhanced the dialectic of representing and intervening, bringing into sharp relief the engineering capabilities of the new biology.

Indeed, in 1949—amidst landmark research on immunology, the biochemical basis of sickle-cell anemia, and the structure of proteins—Pauling announced his long-term vision of engineering life. As a proclaimed molecular architect, he declared his ambition to create life in the laboratory by designing self-reproducing protein molecules. A contemporary article on Pauling's program predicted that "if anyone within the next twenty-five years becomes master of this second creation, it very probably will be Pasadena's wizard of atomic architecture."[17] It was a program that aimed at extending biological control from the lab bench to society; representation became an instrumentality of design and control. It encouraged the mode of thinking inscribed in the foundation's "Science of Man" agenda: one that sought to map the molecular pathways of man's soma and psyche and to rationalize human behavior through molecular knowledge.

The Postwar Era: Life as a Message

When James Watson and Francis Crick elucidated the double-helical structure of DNA and its role in heredity in 1953, a major cognitive hurdle was cleared. The molecular vision of life shifted as the protein paradigm yielded to an emerging paradigm based on DNA, which caused a great reshuffling of ideas and techniques. Max Delbrück correctly prophesied that the new phase would involve structural and analytical chemistry only in part. More important, life scientists would take a fresh look at many problems in genetics and cytology that had come to a dead end during the previous forty years.[18] Equally significant, as the precise mechanisms by which nucleic acids exerted their putative power on organisms were elucidated, molecular biology promised a more precise instrumental rationality through control of the message: control of the DNA sequence. Within the discourse of social control, the so-called "master molecule" was increasingly configured together with other discursive elements of control over life, those derived from cybernetics and information technosciences. Empowered by these new discursive tools, from the 1950s on, molecular biology began to claim even greater cognitive authority and technological prowess in addressing problems of biological deterioration and socioeconomic policy.

To be sure, it is easy to appreciate how the experimental feats of the 1950s could fuel euphoric forecasts and capture the imagination of technological utopianism. Highlights include Joshua Lederberg's construction of the first map of the *E. coli* genome; Matthew Meselson and Franklin Stahl's elegant experiment demonstrating the semiconservative nature of DNA replication; François Jacob and Jacques Monod's elucidation of the mechanisms of bacterial genetic regulation and the role of messenger RNA; and Arthur Kornberg's sensational discovery of DNA polymerase. Though still theoretically limited and technically encumbered, Kornberg's successful execution of an in vitro synthesis of DNA extended the inscriptions of genetic engineering into molecular biology. Alongside these biochemical works, the intensive efforts to represent gene action in terms of a genetic code (completed by 1966), forged formidable links between representing and intervening.[19]

The cognitive, technical, and cultural potency of the genetic code in the development of molecular biology qua genetic engineering is a topic of immense import and scope. The notion of coding and the historical process by which scientists began to speak of organisms and molecules as information storage and retrieval systems and of life as information transfer is part of the legacy of World War II, the Cold War, and the development of command and control systems. The discursive practices and semiotic tools that shaped the notion of a genetic code were not due to the internal cognitive momentum of molecular biology, nor were they the logical outcome of DNA base-pairing. They derived from cybernetics and information theory and were imported into biology, as well as in other disciplines in the life and social sciences, in the late 1940s, still within the protein paradigm of the gene.[20]

The scope of this gestalt shift cannot be reduced to the impact of individual actors, yet looking at key figures does illuminate some of these emergent discursive practices and their relations to molecular biology. The impact of digital control systems on representations of life processes was already both anticipated and promoted by the MIT mathematician Norbert Wiener and his circle in the early 1940s, as they worked on computational problems of ballistics in Warren Weaver's mathematical division within the OSRD (Office of Scientific Research and Development). In 1943, in a noted article, "Behavior, Purpose, and Teleology," Wiener, Julian Bigelow, an engineer trained at MIT, and Arturo Rosenblueth, a Harvard physiologist, first articulated the similarity of servomechanisms, physiological homeostasis, and behavioral processes.[21] Negative feedback as a paradigm of thought and action became Wiener's mission. Even before the war ended, Wiener, the Hungarian émigré mathematician John von Neumann, and their circle began campaigning vigorously for the new field of automated control, yet unnamed.[22]

Von Neumann—by 1945 a key figure in strategic military planning—was then becoming interested in biology in general and in genetics in particular, an interest closely linked to his mission of developing self-reproducing machines. He in fact saw viruses and phage as the simplest models of reproduction. But in linking viruses to information processing von Neumann, like most researchers (especially in the United States), operated within the dominant paradigm in life science. He conceptualized reproduction within the protein view of heredity, in which autocatalytic mechanisms of enzymes explained gene action and virus replication. This is a key point: the introduction of information concepts into genetics predated DNA genetics. He made contacts with the biomedical community, participated in some meetings, and communicated with life scientists, among them Delbrück, Lederberg, and Sol Spiegelman. He was encouraged to develop his self-duplicating machine as a possible heuristic model of gene action.[23]

The cybernetic view of life gained momentum after 1948. That year Wiener's *Cybernetics: Or, Control and Communication in the Animal and the Machine* was published simultaneously in France and the United States. In this remarkably influential book Wiener expounded two central notions: that problems of control and communication engineering were inseparable—communication and control were two sides of the same coin—and that they centered on the fundamental notion of the message—defined as a discrete or continuous sequence of measurable events distributed in time. There he expounded what Foucault would later call a new episteme: the coming of the information epoch. "If the 17th and early 18th centuries are the age of the clocks, and the 18th and 19th centuries constitute the age of steam engines, the present time is the age of communication and control." In this book Wiener explained how the same conclusions held true for both animate and inanimate systems: for enzymes, hormones, neurons, and chromosomes. He was in regular contact with his old friend J. B. S. Haldane, one of his many links to life science. And through these dialogues Wiener's cybernetic view of heredity was grounded in the primacy of proteins. Like von Neumann, Wiener forecast in 1947 that the combinatorial mechanisms by which amino acids organized into protein chains, which in turn formed stable associations with their likes, could well be the mechanisms by which viruses and genes reproduced.[24]

Also in 1948 Claude E. Shannon, an MIT-trained engineer at Bell Laboratories, published a major article, "The Mathematical Theory of Communication," in which he developed the salient features of information theory. His work implied that information can be thought of in a manner entirely divorced from content and subject matter, and he established the basic unit of information as the binary digit, or bit. His highly technical article gained wide exposure through another article he wrote with Warren Weaver, who explained the concepts of information theory in a lucid and eloquent manner.[25]

These ideas found an even wider circulation through Wiener's book *The Human Use of Human Beings,* published in 1950. His thesis: Contemporary society could only be understood through a study of messages and communication facilities; the individual and the organism must be recast in terms of information. In a chapter entitled "The Individual as the Word," Wiener elaborated this concept and its corollary; the technical possibility of writing the messages that represent the organism:

> The earlier accounts of individuality were associated with some sort of identity of matter, whether of the material substance of the animal or the spiritual substance of the human soul. We are forced nowadays to recognize individuality as something which has to do with continuity of pattern, and consequently with something that shares the nature of communication.

It is not matter, but the memory of the form that is perpetuated during cell division and genetic transmission, he insisted. He prophesied that in the future it will be possible to transmit the coded messages that constitute organisms and human beings.[26] We witness here the opening of new discursive space in which the word, or the message, became configured together with the concept of the gene as the locus of technological control, and with notions of ways of controlling bodies that bypassed their materiality.

Although one would not credit Wiener with producing an information discourse, the impact of his ideas was enormous and affected numerous academic fields and cultural sensibilities. I will focus here only on the early diffusion of these ideas beyond the cybernetic group into genetics and molecular biology. Haldane, who invited Wiener to lecture in London, was an enthusiastic convert; he made a sincere and concentrated effort to calculate the information content of various organs and organisms and recast biological accounts in terms of information theory. The geneticist L. S. Penrose was inspired by von Neumann's ideas of self-reproducing automata in the 1950s. And it is virtually certain that when Wiener was invited to lecture at the College de France in 1951, Jacob and Monod attended these lectures. Cybernetics had a wide following in France.[27]

Besides these individual conversions, the Viennese émigré radiologist Henry Quastler undertook a more organized effort to build a new discipline—an information-based biology. Quastler moved from the University of Illinois to Brookhaven Laboratories in the mid-1950s. From 1949 until his untimely death in 1963, Quastler, inspired by Wiener and Shannon and funded through military sources, channeled relentless energy into rewriting biology as an information science. His output—articles, reports, symposia, proceedings, and books—was prolific. In 1952, for example, he organized a major symposium at the Control System Laboratory at the University of Illinois, later published as *Information Theory in Biology*. Two things are striking about the proceedings: the discursive shift to informational representations of life, and the conceptualization of information storage and transfer within the protein paradigm of heredity.

Several participants explained that proteins are especially attractive to information theory. "They are constructed much as a message. . . . [T]he protein molecule could be looked upon as the message and the amino acid residues as the alphabet." In a paper written in 1949 on "error rate and information in living systems," Quastler and Sydney Dancoff established the information content of a human organism to be 5×10^{25} bits. The information content of a single printed page is about 10^4 bits, they noted; thus the description of a human would entail 5×10^{21} pages—a library of roughly ten thousand trillion books. Based on this logic the information content of a germ cell was set at 10^{11} bits; and what they called the "genome catalogue" at the order of about a million bits.[28] A new representation of life phenomena emerged.

These efforts show that a commitment to biological representations in terms of information storage and transfer and its attendant scriptural discourse occurred well before 1953, before DNA replaced proteins as the source of hereditary information. By then a gestalt switch had already occurred. Thus when in 1953 Boris Ephrussi, Urs Leopold, Watson, and J. J. Wiegle wrote a letter to *Nature* about information and bacterial genetics, it did not matter that they were thinking in terms of nucleic acids. They proposed to clean up the semantic confusion in the field by introducing the term "inter-bacterial information." "It does not imply necessarily the transfer of material substances, and recognizes the possible future importance of cybernetics at the bacterial level," they argued.[29] Cybernetics promised a great deal.

A similar discursive shift occurred in the way that the Russian émigré physicist George Gamow represented heredity. In 1954 Gamow introduced the scheme by which a nucleotide triplet specified the assembly of amino acids, a scheme which by 1955 came to be referred to as the genetic code. Inspired by Watson and Crick's discovery and informed by Quastler's efforts, Gamow conceptualized the code as a storehouse of information. When he published the article "Information Transfer in the Living Cell" in *Scientific*

American in 1955, he was already employing familiar discourse—nearly a decade in the making—for he never stopped to justify its usage. The problem, he explained, was one of cell communication—the study of the self-activating transmitter that passes on very precise messages; to understand them we must study the language of cells. Through Gamow and the coding group (Crick, Delbrück, Martynas Yčas, Alexander Rich, and Syndey Brenner, among others), coded messages, genetic texts, and linguistic tropes began to redefine the discursive field of molecular biology in the 1950s; they were initially deployed mainly by physical scientists, but by the 1960s they were adopted by the biochemists as well.[30]

Thus it should be emphasized that when in 1961 Marshall Nirenberg and Heinrich Matthaei at National Institutes of Health "cracked the code," the cognitive and cultural commitments to notions of codes, messages, and texts were already more than a decade old. The demonstration that synthetic polyuridylic acid specified the synthesis of polyphenylalanine showed that the code could be completed by using synthetic chemical probes. From then on amino acids were matched with codons at a staggering rate, especially after a Wisconsin biochemist, Har Gobind Kohrana, managed to synthesize heterogeneous RNA. By 1965 the representation of heterocatalysis in terms of a code was essentially complete: The transcriptional specifications for all the amino acids were on paper.[31] In principle, molecular biology had the apparatus for direct genetic intervention and for designing life in vitro; the proliferating meanings of the genetic code became cultural realities. Tropes such as "the language of cells," "the text of life," "living dictionary" permeated the scientific and popular literature, generating faith that the description of a cell or an organism could be read—and eventually edited—unambiguously from a fixed genomic text. Control of life was to be exercised on the meta-level of controlling the message.

Social Returns: A New Eugenics?

These new inscriptions, discursivities, and instrumentalities produced enormous excitement over potential levels of biological, medical, and social control. In 1958 we find Pauling promoting a technological utopianism grounded in the molecular vision of life in a television broadcast entitled "The Next Hundred Years." Recounting the discoveries relating to biochemical and genetic aspects of disease made in his laboratories, Pauling postulated that there were "thousands, tens of thousands of molecular diseases." In a vision anticipating the human genome initiative, Pauling spoke of the nearing Golden Age, a move away from mere palliative action: biology turning molecular, medicine maturing into an exact science, and social planning becoming rational. Like some of his peers, Pauling saw the deterioration of the human race as the most compelling challenge for the new biology: "It will not be enough just to develop ways of treating the hereditary defects. We shall have to find some way to purify the pool of human germ plasm so that there will not be so many seriously defective children born. . . . We are going to have to institute birth control, population control," he urged.[32]

Pauling was not alone in his molecular utopianism. Inspired by the cascade of findings in molecular biology, in 1963 the Ciba Foundation sponsored a meeting at which some forty distinguished scholars, primarily biologists, gathered to speculate on "man and his future." H. J. Muller promoted, as he had since the 1920s, a new eugenics free of class and race prejudices and based on biological and social merit. "It is more economical in the end to have developmental and physiological improvements of the organism placed on a genetic basis, where practicable, than to have to institute them in every generation anew by elaborate treatment of the soma," he argued. Lederberg, though not in full agreement with Muller, predicted that in "no more than a decade the molecular knowledge of microbes would be applied to the human genome." As he saw it, "the ultimate application of molecular biology would be the direct control of nucleotide sequences in human chromosomes, coupled with recognition, selection, and integration of the desired genes, of which the existing population furnishes a considerable variety. These notions of a future eugenics are, I think, the popular view of the distant role of molecular biology in human evolution." Crick too agreed generally with these goals, questioning only the feasibility of their implementation.[33]

In the 1960s genetic engineering was seen as an emergent medical and social technology. Bud has described how Tatum, who had urged researchers to think about the possibilities of biological engineering in the late 1950s, championed genetic engineering in the 1960s. Even as public alarm and mistrust of the new technologies began to gather momentum, he promoted genetic engineering as a three-tier project: recombination of existing genes, or eugenics; the production of new genes through directed mutations, or genetic engineering; and modification or control of gene expression, or in Lederberg's term, *euphenic engineering.* At the inauguration of the laboratories of the pharmaceutical conglomerate Merck, Sharp and Dohme, Tatum spiced up his message with a provocative ditty.

The time has come, it may be said,
To dream of many things
Of genes—and life—and human cells
Of Medicine—and Kings.[34]

Around the same period of public debate and alarm, Pauling proposed a kind of "yellow star" policy of eugenic prophylaxis: "There should be tattooed on the forehead of every young person a symbol showing possession of the sickle-cell gene or whatever other similar gene. . . . It is my opinion that legislation along this line, compulsory testing for defective gene before marriage, and some form of semi-public display of this possession, should be adopted."[35] And at Caltech, Robert Sinsheimer rejoiced in the new powerful technologies for perfecting the remarkable product of two billion years of evolution: "The old eugenics was limited to a numerical enhancement of the best of our existing gene pool. The new eugenics would permit in principle the conversion of all the unfit to the highest genetic level."[36]

These euphoric forecasts were not politically neutral, nor were they distant visions or wishful fancy. Politically loaded, they conveyed hard line positions and represented active campaigns in an escalating national debate over the control of animate nature. It was the beginning of a political struggle over the representation of and intervention in life processes, and about the appropriation of disciplinary symbolic capital: genetic technologies. In the mid-1960s a counterdiscourse emerged that in the following decades seriously challenged the hegemony of the molecular vision of life. Charles Weiner has explored the way that discussions of genetic engineering spilled over from scientific meetings into the domains of Congress and the media, so that several biologists feared loss of control over the debates. The public alarm over an impending genetic engineering forced molecular biologists to confront the ethical issues with different degrees of commitment. For Salvador Luria the ethical issues of engineering humans by deploying knowledge garnered from viruses and bacteria were secondary—in priority and significance—to issues of technical feasibility. Marshall Nirenberg, on the other hand, placed the ethics before technical issues. Humans must wait for sufficient wisdom before attempting to instruct their own cells, he believed.[37] These debates exhibit the immediacy and urgency that genetic engineering acquired in the 1960s and foreshadow the more intense confrontations in the 1970s and 1980s over the use of recombinant DNA technologies.

Conclusion

How do we situate this molecular technological utopianism and the nascent public alarm in response to it? Were the molecular biology researches in the 1950s and 1960s in any sense mission-oriented? Were they carried along by the social momentum generated since the 1930s by the goal-directed agenda of the Rockefeller Foundation? Were there ideological continuities? Not only were the political and institutional configurations of postwar science complex, but there are few studies of postwar biology to build on. It is hard at present to assess clearly the social forces behind various research programs in molecular biology after the mid-1950s. As the life sciences became decentralized, molecular biology enjoyed a pluralistic mode of patronage, one that relied on the government, the military, industry, and private foundations. The scientific ideology of social control, so neatly articulated and localized by the ruling academic-business elite of the 1920s, was fragmented and lodged within pockets of power dispersed throughout the multiple administrative structures of postwar science.

The articulations of biopower also changed by the 1960s. New modes of signification, those derived from information technosciences, came to empower the molecular vision of life. Within a new economy of discourse, the soma, psyche, normalcy and deviance, genomes, nucleic acids, proteins, information, texts, and messages were systematically configured together. Beyond the material level, the control of bodies and populations now hinged, as Lederberg projected, on the direct control of the sequence, on editing and rewriting the organismic text.

Yet the lines of ideological continuity from the 1920s to the 1960s are as remarkable as these institutional and cognitive discontinuities. The underlying epistemological commitment of molecular biology survived the paradigm shift from a protein base to a DNA base—and survived the change in patronage. The premise that the soma and psyche are essentially the outcome of the genetically determined activity of macromolecules, and that these mechanisms of upward causation should be the principal basis for intervening in higher-order life processes, has acquired even greater intellectual vigor and social legitimacy, braced by institutional and commercial interests that dwarf the millions of dollars of the Rockefeller Foundation. And, as in the 1930s, the sweeping postwar expressions of faith in technologies of selective breeding attest to the durability, resilience, and lure of a science-based social control and the quest for biopower.

As I have argued, molecular biology in the pre-recombinant DNA era was not a pure science, cloistered in the academy, with little relevance to the marketplace. Molecular life as a commodity did not abruptly spring up in the 1970s, like Athena from the head of Zeus. The flurry of commercial and public attention in the 1970s and 1980s certainly created a sense, or sensationalism, of novelty in the academy and the boardroom. Yet below such surface activity the lines of continuity reach back to the origins of the molecular biology program of the 1930s and account equally well for the field's broader market value and commodification.

In order to extract the different levels of significance inherent in these continuities, I have problematized the category of basic research and the perceived movement from the pure to the applied. I have argued that the expectations and articulations of social and biological engineering were written into the molecular biology program from its inception, as inscriptions, discursivities, and instrumentalities. Articulated within the historically specific and culturally contingent categories of social control of the 1930s, with their attendant discourse, the program was justified and supported for its long-term market value. Within the framework of biopower—the control of bodies and the control of populations—life came to be represented as a genetically directed activity of macromolecules. The discursive shifts shaped by the historical specificities of the 1950s—the linguistic and semiotic tools of the information technosciences—did modify that discourse to include a meta-level nonmaterial dimension of animate control: control of the message. And the discontinuities in the patronage of molecular biology clearly had an important impact on the organization of the field in the postwar era. Yet molecular biology as a technology-driven and technology-producing program continued to embody a dialectic of representing and intervening. And commodification, in principle or in practice, was not very far from the laboratory bench.

Notes

1. Robert Bud, *The Uses of Life: A History of Biotechnology* (Cambridge: Cambridge University Press, 1993), pp. 1 (definition given by the Organization for Economic Co-operation and Development), 213 (NSF-FDA comment).

2. Bud, *Uses of Life* (note 1), chap. 2.

3. Ian Hacking, *Representing and Intervening: Introductory Topics in the Philosophy of Natural Science* (Cambridge: Cambridge University Press, 1983), pp. 130–46. On the interventionist nature of the Baconian program see Carolyn Merchant, *The Death of Nature: Women, Ecology, and the Scientific Revolution* (San Francisco: Harper and Row, 1980), chaps. 7–9.

4. Ian Hacking, "Weapons Research and the Form of Scientific Knowledge," *Canadian Journal of Philosophy* Supp. 12 (1986), 235–50; Paul Forman, "Behind Quantum Electronics: National Security as Basis for Physical Research in the United States, 1940–1960," *Historical Studies in the Physical and Biological Sciences* 18 (1987), 149–229; Evelyn Fox Keller, "Critical Silences in Scientific Discourse: Problems of Form and Re-Form," in Keller, *Secrets of Life, Secrets of Death: Essays on Language, Gender and Science* (New York: Routledge, 1992), pp. 73–92; and Keller, "Physics and the Emergence of Molecular Biology: A History of Cognitive and Political Synergy," *Journal of the History of Biology* 23 (1990), 389–410. For a related argument see Philip Pauly, *Controlling Life: Jacques Loeb and the Engineering Ideal in Biology* (New York: Oxford University Press, 1987).

5. Robert E. Kohler, *Partners in Science: Foundations and Natural Scientists, 1900–1945* (Chicago: University of Chicago Press, 1991); Kohler, "The Management of Science: The Experience of Warren Weaver and the Rockefeller Foundation's Programme in Molecular Biology," *Minerva* 14 (1976), 249–93; Edward J. Yoxen, "Giving Life a New Meaning: The Rise of the Molecular Biology Establishment," in *Scientific Establishments and Hierarchies: Sociology of the Sciences,* ed. Norbert Elias, Herminio Martins, and Richard Whitley (Dordrecht: Reidel, 1982), pp. 123–43; Pnina Abir-Am, "The Discourse of Physical Power and Biological Knowledge in the 1930s: A Reappraisal of the Rockefeller Foundation's 'Policy' in Molecular Biology," *Social Studies of Science* 12 (1982), 341–82; and Lily E. Kay, *The Molecular Vision of Life: Caltech, the Rockefeller Foundation, and the Rise of the New Biology* (New York: Oxford University Press, 1993).

6. This argument is elaborated in Kay, *Molecular Vision of Life* (note 5), chap. 1.

7. Raymond B. Fosdick, *The Old Savage and the New Civilization* (New York: Doubleday, Doran, 1928), pp. 21–31, 184, on p. 31.

8. Weaver's report, 14 Feb. 1934, pp. 2–3, Rockefeller Archive Center, record group 3, box 1.7; excerpts in Raymond B. Fosdick, *The Story of the Rockefeller Foundation* (New York: Harper, 1952), p. 158.

9. The original concepts of cultural hegemony and the role of intellectuals in society were formulated by Antonio Gramsci, *Selections from the Prison Notebooks* (1929–1935), ed. and trans. Quintin Hoare and Geoffrey Nowell Smith (New York: International Publishers, 1971). For an elaboration of the Gramscian framework and its use in American history see T. J. Jackson Lears, "The Concept of Cultural Hegemony: Problems and Possibilities," *American Historical Review* 90 (1985), 567–93, and "A Round Table: Labor, Historical Pessimism, and Hegemony" (six papers on the scope and limits of Gramscian hegemony), *Journal of American History* 75 (1988), 115–62. For an excellent analysis of the concept of cultural hegemony see also Jean Comaroff and John Comaroff, *Of Revelation and Revolution: Christianity, Colonialism, and Consciousness in South Africa,* vol. 1 (Chicago: University of Chicago Press, 1991), intro.

10. Kay, *Molecular Vision of Life* (note 5), intro., chap. 1.

11. Michel Foucault, *The History of Sexuality,* vol. 1: *An Introduction,* trans. Robert Hurley (New York: Pantheon, 1978); and Foucault, *Power/Knowledge: Selected Interviews and Other Writings, 1972–1977,* ed. Colin Gordon (New York: Pantheon, 1980), pp. 92–133. For an excellent discussion on the role of discursive formations in the emergence of disciplines, see Timothy Lenoir, *Instituting Science* (Palo Alto, Calif.: Stanford University Press, forthcoming), chap. 2: "The Discipline of Nature and the Nature of Disciplines."

12. For a discussion on cooperation as an organizational, managerial, academic, behavioral, and ideological category see Kay, *Molecular Vision of Life* (note 5), chaps. 1, 2; on the relation to biological research see "Interlude I."

13. Joseph S. Fruton, *Molecules and Life* (New York: Wiley and Sons, 1972), chap. 3; Robert C. Olby, *The Path to the Double Helix* (London: Macmillan, 1978), chap. 2; and Norman H. Horowitz, "Genetics and the Synthesis of Proteins," in *Origins of Modern Biochemistry: A Retrospect on Proteins,* ed. P. R. Srinivasan, Joseph S. Fruton, and John T. Edsall, Annals of the New York Academy of Sciences 325 (New York: New York Academy of Sciences, 1979), 253–66.

14. Lily E. Kay, "Selling Pure Science in Wartime: The Biochemical Genetics of G. W. Beadle," *Journal of the History of Biology* 22 (1989), 73–101.

15. Lily E. Kay, "Molecular Biology and Pauling's Immunochemistry: A Neglected Dimension," *History and Philosophy of the Life Sciences* 11 (1989), 211–19; and Kay, *Molecular Vision of Life* (note 5), chap. 6.

16. Kay, *Molecular Vision of Life* (note 5), chap. 6.

17. "Linus Pauling, Atomic Architect," *Science Illustrated* (Jan. 1949), 39; Kay, *Molecular Vision of Life* (note 5), pp. 261–62; Kay, "Life as Technology: Representing, Intervening and Molecularizing," *Rivista di Storia della Scienza* Ser. II, 1:1 (June 1993), 85–103.

18. Delbrück papers, box 23.20, Delbrück to Watson, April 14, 1953, California Institute of Technology Archives.

19. On these developments see Horace Freeland Judson, *The Eighth Day of Creation: The Makers of the Revolution in Biology* (New York: Simon and Schuster, 1979); and Gunther S. Stent and Richard Calendar, *Molecular Genetics: An Introductory Narrative* (San Francisco: W. H. Freeman, 1978). To date there is no single critical history of the genetic code. Among the useful works on various dimensions of the genetic code and the history of molecular biology, the oldest and most detailed is Judson's *Eighth Day of Creation,* but it has little critical distance from the actors. More recent treatments of the subject include Richard Doyle, "On Beyond Living: Vital and Post-Vital Rhetorics in Molecular Biology" (Ph.D. dissertation, University of California, Berkeley, 1993); Daniel J. Kevles and Leroy Hood, eds., *The Code of Codes: Scientific and Social Issues in the Human Genome Project* (Cambridge, Mass.: Harvard University Press, 1992); Evelyn Fox Keller, "Secrets of Life, Secrets of Death," *New York Review of Books,* 4 June 1992, pp. 31–40; and on postwar molecular biology more generally, Pnina

Abir-Am, "The Politics of Macromolecules: Molecular Biologists, Biochemists, and Rhetoric," *Osiris* 7 (1992), 164–94. I am currently working on a history of the genetic code.

20. The origins of the genetic code have been attributed to Erwin Schrödinger—a view promoted by Stent and Calendar, *Molecular Genetics;* propagated by Olby, *The Path to the Double Helix* (note 13) and by Judson, *The Eighth Day of Creation;* and tacitly accepted by Richard Doyle, "Mr. Schrödinger Inside Himself," *Qui Parle* 5 (1992), 1–20. I believe that the influence of information technoscience and military technologies are of greater importance. See also Lily E. Kay, "Who Wrote the Book of Life? Information and the Transformation of Molecular Biology," in *Objekt, Differenzen, und Konjunkturen: Experimentalsysteme in historischen Kontext,* ed. Michael Hagner, Hans-Jörg Rheinberger, and Bettina Wahring-Schmidt (Berlin: Akademie Verlag, 1994), pp. 151–79.

21. Arturo Rosenblueth, Norbert Wiener, and Julian Bigelow, "Behavior, Purpose, and Teleology," *Philosophy of Science* 10 (1943), 18–24. See also Peter Galison, "The Ontology of the Enemy: Norbert Wiener and the Cybernetic Vision," *Critical Inquiry* 21 (1994), 228–66.

22. Norbert Wiener to John von Neumann, 17 Oct. 1944; and von Neumann to Wiener, 16 Dec. 1944, Massachusetts Institute of Technology Archives, Wiener papers MC 22, box 2.66.

23. Von Neumann to Wiener, 29 Nov. 1946, ibid., box 2.72; Library of Congress, von Neumann papers, box 7.1, Spiegelman to von Neumann, 3 Dec. 1949; and von Neumann to Spiegelman, 10 Dec. 1946. See also John von Neumann, "The General and Logical Theory of Automata," in *Cerebral Mechanisms in Behavior: The Hixon Symposium* (New York: Hafner, 1951), pp. 1–32; Steve J. Heims, *John von Neumann and Norbert Wiener: From Mathematics to the Technologies of Life and Death* (Cambridge, Mass.: MIT Press, 1980); and William Aspray, *John von Neumann and the Origins of Modern Computing* (Cambridge, Mass.: MIT Press, 1990).

24. Norbert Wiener, *Cybernetics: Or, Control and Communication in the Animal and the Machine,* 2nd ed. (New York: MIT Press and John Wiley and Sons, 1961), intro. (quotation) and pp. 93–94. See also Michel Foucault, *The Order of Things: An Archaeology of the Human Sciences* (New York: Vintage Books, 1970).

25. Claude E. Shannon, "A Mathematical Theory of Communication," *Bell Systems Technical Journal* 27 (1948), 379–423, 623–56; and Shannon and Warren Weaver, *The Mathematical Theory of Communication* (Urbana: University of Illinois Press, 1949).

26. Norbert Wiener, *The Human Use of Human Beings* (Boston: Houghton Mifflin, 1950), pp. 103, 108–11, on p. 103.

27. Kay, "Who Wrote the Book of Life?" (note 20).

28. See articles in Henry Quastler, ed., *Essays on the Use of Information Theory in Biology* (Urbana: University of Illinois Press, 1953), e.g., Herman R. Branson, "Information Theory and the Structure of Proteins," pp. 84–104, on pp. 84–85; and Sydney Dancoff and Henry Quanstler, "Information Content and Error Rate in Living Things," pp. 263–73.

29. B. Ephrussi, Urs Leopold, J. D. Watson, and J. J. Wiegle, "Terminology in Bacterial Genetics," *Nature* 171 (1953), 701.

30. George Gamow, "Information Transfer in the Living Cell," *Scientific American* 193 (April 1955), 70–78. The diffusion into the biochemical community is discussed by Lily E. Kay, "Matter of Information: Changing Meanings of the Tobacco Mosaic Virus, 1930s–1960s," paper presented at the XIXth International Congress of History of Science, Zaragoza, Spain, August 1993.

31. Judson, *Eighth Day of Creation* (note 19), part 2.

32. Pauling file, "The Next Hundred Years," KRCA, Channel 4, 13 Dec. 1958, historical file, box 88, California Institute of Technology Archives.

33. Hermann J. Muller, "Genetic Progress by Voluntarily Conducted Germinal Choice," in *Man and His Future: A Ciba Volume,* ed. Gordon Wolstenholme (Boston: Little, Brown, 1963), pp. 254–56; Joshua Lederberg, "Biological Future of Man," ibid., pp. 264–65 (Lederberg also advocated euphemic engineering, or controlling modes of gene expression); and Crick in "Eugenics and Genetics" (discussion), ibid., pp. 286–89.

34. Edward L. Tatum, "Molecular Biology, Nucleic Acids, and the Future of Medicine," *Perspectives in Biology and Medicine* 10 (1966–67), 31; and Tatum, "Perspectives from Physiological Genetics," in *The Control of Human Heredity and Evolution,* ed. T. M. Sonneborn (New York: Macmillan, 1965), p. 22 (the poem); both quoted in Bud, *Uses of Life* (note 1), p. 170.

35. Linus Pauling, "Reflections on the New Biology," *UCLA Law Review* 15 (1968), 269, quoted in Troy Duster, *Backdoor to Eugenics* (New York: Routledge, 1990), p. 46.

36. Robert Sinsheimer, "The Prospect for Designed Genetic Change," *Engineering and Science* 32 (1969), 8–13, quoted in Evelyn Fox Keller, "Nature, Nurture, and the Human Genome," in *Code of Codes,* ed. Kevles and Hood (note 19), pp. 289–90.

37. Charles Weiner, "Anticipating the Consequences of Genetic Engineering: Past, Present, and Future," in *Are Genes Us? Essays on the Ethical and Social Consequences of Genetics and the Human Genome Initiative,* ed. Carl F. Cranor (New Brunswick, N. J.: Rutgers University Press, 1994).

Biotechnology and Blood: Edwin Cohn's Plasma Fractionation Project, 1940–1953

Angela N. H. Creager

The emerging power of biotechnology in contemporary society is most commonly ascribed to two factors: the conjunction of public knowledge and private gain in university-industry alliances and the development in the 1970s of techniques for genetic engineering. The word *biotechnology* itself, however, came into currency in several national contexts long before the invention of cloning techniques.[1] Indeed, some scholars and champions of biotechnology have even pointed to Louis Pasteur as the founder of biotechnology in modern Western society. Nonetheless, much recent scholarship has attributed the expansion of biotechnology to the scientific and technological advances made during the 1970s in manipulating recombinant DNA. Accordingly, many historians and science writers have organized the history of biotechnology around DNA rather than around altered relations between the laboratory and the market, even as they stress the changes in these relations occasioned by the industrialization of genetic engineering.[2] Like the articles by Lily Kay and Robert Bud, my contribution to this interdisciplinary volume on the development of biotechnology focuses on the earlier history of entrepreneurial technology within "basic" molecular biology and biochemistry.

From this historical perspective, what is new and striking about the conversion of genetic experimentation in the 1970s into today's corporate biotechnology is the way molecular biologists' research practices were taken to be authoritative and economically significant, not simply laboratory-bound scientific feats. Claims about the power of molecular biological techniques to support a new industry of biologics were readily accepted despite the slow pace at which commercially viable products have come to market (with the dramatic exception of insulin). The perceived authority of laboratory scientists among U.S. corporations, the federal government, investors, and the public was not a response to genetic engineering, but rather a precondition to its commercialization. My aim in this article is to show how the stage was set for the reception of the scientific innovations associated with recombinant DNA. I shall do this by exploring one central episode in the mobilization of biochemistry during World War II and the subsequent industrial outcomes.

The Plasma Fractionation Project (PFP), headed by Edwin Cohn of Harvard Medical School, developed and produced novel therapeutics from human blood beginning in 1940 (with the newly formed national blood donation service of the Red Cross providing starting material). During World War II, the PFP involved university laboratories at Wisconsin, Stanford, and Columbia as well as Cohn's Harvard Medical School laboratory (the focus of this essay); clinical testing units in hospitals in New York, Philadelphia, Richmond, Atlanta, Boston, and San Francisco; and seven pharmaceutical companies throughout the country. The work at all of the sites was directed and coordinated by Edwin Cohn, who continued the collaboration between his laboratory and the pharmaceutical industry after the war. The emergence of commercial blood-derived therapeutics from these sources provides an important although problematic precedent for contemporary biotechnology in several regards.

First, the alliance between Cohn's Harvard Medical School laboratory and several pharmaceutical companies, which privileged academic expertise, presages the links between "basic" university molecular biologists and biotechnology companies that emerged in the 1970s and 1980s. In each case, commercialization of a biological object required replicating and standardizing a laboratory method, and scaling it up dramatically. The transfer of knowledge from the bench of a university investigator to an industrial plant requires multiple negotiations and adjustments to keep the authority of the innovator intact. Some companies have become expert at navigating this transition, and several of the pharmaceutical companies who participated in the PFP are numbered among the large corporations now committed to genetically engineered drugs.

Second, the Plasma Fractionation Project resembles recent biotechnology in the way that it "altered a whole set of symbolic and cultural relationships that people have with their bodies."[3] Like the bioproduction of insulin and human growth hormone, the production of pharmaceuticals from blood occasioned a new conception of the human body and a fragmenting of the distinction between its "natural" and synthetic components. Whereas the therapeutics of the PFP were obtained from donated human blood, many new pharmaceuticals are derived from microorganisms expressing a human protein or hormone from a re-engineered fragment of the human genome. Transfusion technologies elevated the status of the biochemistry laboratory as the universalizer of human body constituents, a perception crucial to the success of subsequent biomedical interventions, from organ transplantation to gene therapy. The history of the medical uses of biotechnology could be written as a history of the fractionation of the body in the laboratory, and although I only allude to this important story here, the PFP occupies a significant place in that account.[4]

Third, the boundaries that the contemporary biotechnological enterprises are frequently perceived to transgress, such as that between public knowledge and private profit, were (re)constituted in large measure as a result of the scientific mobilization effort of World War II.[5] The PFP exemplified the success of war-related scientific projects used to justify the expansion of federal funding for research after the war. Yet the legacy of World War II, as illustrated by the PFP, was one not of pure but of applied research. "Basic" federally funded research was a category constructed to continue laboratory activity and authority after mobilization; it was part of how civilian life was reconstituted after the war in conjunction with "its inseparable twin, the military."[6] Although Cohn's success was touted as the triumph of "basic" research expertise applied to a practical problem, the economic consequences of the war project made it doubtful that Cohn would return to administering academic research alone after the war. Martin Kenney has noted that "in contracting with private companies to produce blood plasma, the government simultaneously created a new commodity."[7]

Cohn endeavored to continue his role in the production of plasma-derived commodities just when norms for scientific production emerging after the war amplified the perceived distance between pure biological research and industrial production. The expansion of the National Institutes of Health (NIH) extramural grants program and the founding of the National Science Foundation (NSF), with substantial funding for basic research in biology, enabled biochemistry and molecular biology to expand without increased dependence on either private philanthropic support or industry.[8] Cohn was exceptional among postwar biomedical scientists at prestigious institutions for continuing his work with industry when there was ample funding to return to academic research. In 1953, the PFP was judged to be "big business" in a university setting, an impropriety resembling the later transgressions of the public-private divide often attributed to biotechnology.

Both the PFP and current biotechnological enterprises stand in a longer history of collaborations between industry (here the pharmaceutical industry) and university laboratories in the United States. As John Swann has argued, "cooperative biomedical research between universities and industry in America emerged as a general movement between the two world wars."[9] Such collaborations took on a variety of forms, and opinion regarding the use of patents by university researchers varied over time among the public and in the medical community.[10] The PFP represents the culmination of this interwar trend of pharmaceutical company-university collaborations, but the exigencies of war and the consequent involvement of the government and the military set it apart from earlier cooperative ventures. Moreover, the aftereffects of the mobilization changed the relations between university researchers and pharmaceutical companies in complicated ways. While war research brought together university scientists and industrial producers, it also inaugurated large-scale federal funding of biomedical research, freeing prominent university biologists in the 1950s through the 1970s from having to collaborate with industry for support. Contacts between academic scientists and industry weakened after World War II, reaching their nadir in the 1970s.[11] In the past two decades, academic researchers and industries have once again collaborated in efforts to commodify genetically engineered objects as foods and drugs. The fate of Edwin Cohn and the PFP reminds us of the transitory nature of the status conferred upon a "pure" researcher once commercial ties are established.

University and Industry: Related by Blood

In his foreword to *Scientists Against Time*, a 1947 account of the research conducted under the Office of Scientific Research and Development, Vannevar Bush praised the patriotism and technical achievements of the scientists who worked on the war effort. In his nearly jingoistic account of science in the war effort, one sentence is unusually profound: "This is the story of the development of weapons of war, but it is also the story of an advance in the whole complex of human relations in a free society, and the latter is of the greater significance."[12] The Plasma Fractionation Project, centered at the Department of Physical Chemistry at Harvard Medical School, exemplifies his insight. The linkages established between industrial producers, the university laboratory, and the clinic as science was mobilized for war lasted beyond 1945 to shape postwar civilian medical practice. The emergence in the United States of blood collection, separation, and distribution as an industry relied on the cooperation during World War II of academic protein biochemists, clinicians, the military, pharmaceutical houses, and the millions of civilians who provided the material for the production of novel biologics.

Transfusion, the technology of moving blood from one human body to another, became a reliable medical practice early in the twentieth century. The identification of the A, B, and O immunological blood groups at the turn of the century, and the later construction of rhesus positive and negative groupings, provided a scheme for classifying donors and recipients to protect against adverse reactions to blood from another body. War accelerated the improvement of transfusion techniques, "paving the way for the general adoption of blood transfusion as a measure of proved therapeutic value."[13] The development of an anticoagulant citrate solution in 1915 enabled blood to be collected at one time, stored, and then transfused later, an advance that fundamentally altered transfusion practice. World War I saw the first uses of stored blood by military surgeons, and storage of whole blood extended widely the applicability of transfusion for both civilian and military medicine. Significant advances were made in the practice of blood transfusion in the 1920s and 1930s, with the first blood bank in the United States opening at Cook County Hospital in Chicago in 1937. Also in the 1930s, the Republican Army in the Spanish civil war was using a blood transfusion service, the first of its size in a military context.[14]

Military medicine relied on laboratory-based technologies, such as standardized blood-typing techniques, not only for the widespread adoption of whole blood transfusions, but for the provision of alternatives in the form of blood substitutes. Whole blood can only be preserved under sterile and refrigerated conditions; even under these optimum circumstances, red blood cells do not last for longer than three weeks. During World War I, the inability of the American military to overcome these limitations had led to the use of other liquids in transfusion. Shock was thought to result from the depletion of blood volume, which could be replaced with nonphysiological solutions as well as by blood or its derivatives. As related by two prominent transfusion researchers,

> The basis for the use of blood substitutes in the emergency treatment of shock stems from two fundamental physiological concepts: first, that wound shock is the result of rapid diminution in blood volume rather than loss of hemoglobin; and second, that the efficacy of any replacement fluid depends on its molecular properties, which determine the osmotic pressure it exerts and the extent to which it is retained within the bloodstream. These concepts underlay the introduction of the colloid, gum acacia, during World War I.[15]

A similar rationale led to the use of blood derivatives rather than whole blood to restore blood volume to casualties. Plasma, the liquid part of blood without its corpuscles, became a common transfusion liquid by World War II, in part because it does not require donor-recipient cross matching. Another advantage of plasma was its longevity when dried, and the recent invention of freeze-drying machines, or *lyophilizers*, by Sharp and Dohme, made it possible to stockpile dried plasma.

In the spring of 1940, as part of the preparedness period, the U.S. armed forces had decided to stockpile blood substitutes for transfusion because the technologies for whole blood storage (particularly refrigeration) were limited. Dried plasma was favored as a therapeutic for shock in lieu of whole blood, but it was challenging to amass because large numbers of blood donors were required. The navy hoped that a blood substitute could be developed from bovine blood, which was readily available from Midwestern slaughterhouses (bovine plasma being too antigenic to employ directly as a blood substitute). This perceived need for new blood substitutes provided a research agenda for the participation of university researchers, especially protein chemists such as Cohn, in the war effort.[16]

Also critical to transfusion medicine during the war was a system of production larger than that the university laboratories or hospitals could provide. As the United States mobilized to supply transfusion materials to its allies and then to its own troops in 1940 and 1941, the cooperation of the pharmaceutical houses in producing blood-derived biologics was essential. Precedents in the use of pharmaceutical corporations to meet public and military needs included the transfer of antitoxin production from Public Health Laboratories to commercial houses, and the large government contracts given pharmaceutical companies during World War I, when antitoxin and Salvarsan were important therapeutics.[17] The success of the PFP relied on the coordination of university researchers and industrial producers through the military, and this conjunction, like that enabling the production of antibiotics, would be celebrated for its postwar consequences.[18] In 1945, A.N. Richards, chairman of the Committee on Medical Research of the Office of Scientific Research and Development, spoke "to a fascinated Senate committee on the breakthrough in blood plasma [fractionation]": "That investigation was started at Harvard University, with support of the Rockefeller Foundation, back in 1938, and has been continued up to the present. . . . At the beginning, it was on the laboratory scale, wholly, and the products were curiosities. Now the products are commercially available and being supplied, or are ready to be supplied, in huge quantities."[19] The story of how laboratory "curiosities" became commercially available products is the classic narrative of biotechnology. Although collaborations between universities and the pharmaceutical industry were not uncommon prior to World War II, the demand for stable transfusion materials and medical supplies expanded and strengthened these connections, with Cohn's precedent-setting role in industry continuing after the military sponsorship ceased.[20]

Contingencies and Collaborations in World War I

In the absence of war, the development of transfusion technologies would have almost certainly continued along the lines of whole blood transfusion, obverting the immediate need for new biochemical expertise and pharmaceutical production. In fact, by the end of World War II, many regarded as a mistake the almost total reliance early in the war on such blood substitutes as stockpiled dried plasma and serum albumin.[21] Nonetheless, these products had become firmly entrenched in transfusion practice, and serum albumin continued to be mass produced after the war. In addition, the "by-products" of serum albumin production — other plasma proteins — were found therapeutically useful, which reinforced the demand for blood-derived biologics. As the consequences of Cohn's war project for the pharmaceutical industry reflected changes in the relationship between university researchers and the federal government, a foray into this background is warranted.

The mobilization for war in 1940 changed the landscape for research support in the life sciences as it did in the physical sciences. Basic biology research in the previous decades had been sponsored largely by private foundations, most notably the Rockefeller Foundation.[22] Pharmaceutical companies that collaborated with university researchers in biomedical research constituted another source for research support.[23] The relationship between scientists and the federal government was not based principally on research patronage. Scientists had aided the government's efforts to develop technologies and supplies for World War I through the National Research Council (NRC), founded through the National Academy of Sciences in 1915.[24] The NRC remained a permanent body for supplying the government with scientific advice, but the National Academy of Sciences opposed dependence of scientists on the federal government for funding as a matter of principle.[25] This policy of independence began to erode under the financial crises of the 1930s, when the federal government supported university research through the Works Progress Administration.[26] Yet even then, most government money was directed toward applied research. The impending war in Europe made fundamental research more attractive to the federal government and the military, because both perceived an urgent need for new laboratory-based technologies. World War II allowed laboratory scientists to achieve new status by addressing national needs.

Vannevar Bush believed that scientific research could significantly contribute to U. S. preparedness for war, and in the summer of 1940 he organized the National Defense Research Committee (NDRC) with the approval of President Roosevelt.[27] The NDRC was authorized to organize and fund war-related research conducted by educational institutions, scientific organizations, individuals, and industries. A year later this funding strategy expanded into the Office of Scientific Research and Development (OSRD), under the directorship of Bush, to accommodate a more comprehensive system of contracts for war research. Although war-related research was by necessity applied, the organization of both the NDRC and the OSRD permitted individual initiative and provided generous funding for scientists willing to develop new weapons and other technologies for war.

The NDRC at its founding was composed of two representatives from the military, the commissioner of patents, and five prominent scientists. One scientist, Frank Jewett, was president of both the National Academy of Sciences and Bell Telephone Laboratories; and the other scientists were also highly regarded administrators of institutions known for their scientific research. Bush presided over the Carnegie Institution of Washington; Karl Compton, a noted physicist, was president of the Massachusetts Institute of Technology; James Conant, a chemist, was president of Harvard University; and Richard Tolman, a physicist, was dean of the graduate school at the California Institute of Technology. This quorum of scientific elites, rather than establishing national war laboratories, organized the war research efforts around contracts to top scientists at universities, industries, and institutes. This decision to fund war research by contract to preexisting laboratories, university and industrial, set an important precedent in the relationship of the U.S. government to laboratory research. In addition, the development of weapons and other new military technologies outside military laboratories gave the research scientists, particularly those working in university settings, a new authority in matters of national defense.

The way in which the precedent of OSRD and defense contracts reorganized U.S. research in the physical sciences has been amply explored by historians of science. The developments of the microwave radar, the ENIAC computer, the proximity fuse, and the atom bomb not only transformed the character of the war, but altered the course of the physical sciences, as has been well documented for physics.[28] The mobilization of science for World War II had a profound impact on the life sciences also, but the institutional framework was somewhat different. The original organization of the OSRD did not include biomedical research, but early in the preparedness period the military enlisted scientific expertise relevant to medical problems through other channels.

In the spring of 1940, the NRC organized two committees of scientists at the request of the Surgeon General to provide advice on expected military medical needs in transfusion and chemotherapy.[29] A year later, in June 1941, the Committee on Medical Research (CMR) was constituted as a primary subdivision of the OSRD. The CMR began to fund the research needed by the NRC committees, as well as biomedical investigations in many other areas. The most generously funded CMR program concerned malaria research; aviation medicine and the production of penicillin were next; and the program for research on blood substitutes for transfusion was fourth. Other sponsored research included projects on infectious diseases, venereal diseases, neuropsychiatry, wounds and burns, and adrenocortical hormones.[30] Research on nutrition was also generously supported by the CMR, and it was through this avenue that the work of George Beadle and Edward Tatum on biochemical genetics, which contributed profoundly to postwar molecular biology, was funded.[31]

On 31 May 1940, the NRC's Committee on Transfusions met for the first time to discuss donating transfusion units to the allied troops on the Continent. Representatives from the American Red Cross were consulted, and the first blood donation services of the Red Cross were planned.[32] Harvard Medical School was an institutional nexus for several of the scientists who participated in the effort to establish the transfusion materials infrastructure. The chairman of the medical advisory committee of the American Red Cross was David Edsall, former dean of the Medical School. The distinguished Harvard physiologist Walter Cannon served as chairman of the

NRC Committee on Transfusion and also served on the OSRD-CMR with A. Baird Hastings, the chairman of the Department of Biological Chemistry at Harvard Medical School. When "Cannon, representing the National Research Council Committee on Shock [sic], [asked] who should undertake to examine the use of beef blood proteins as human plasma expanders," Hastings replied, "Edwin J. Cohn, of course." Hastings supervised the disbursement of funds to Cohn's laboratory throughout the war, when "Cohn's contract with the CMR ran into the hundreds of thousands of dollars a year."[33]

Edwin J. Cohn headed the Department of Physical Chemistry at Harvard Medical School, a research department sustained throughout the 1930s by generous grants from the Rockefeller Foundation to investigate questions of protein structure.[34] Cohn was not a clinical researcher and did not teach the medical students at Harvard. His sole venture into medical research was the isolation of a liver extract in 1927 for treatment of pernicious anemia.[35] He was nonetheless a very prominent biochemist who built a protein research program incorporating the most recent theories in physical chemistry. Cohn's studies on the electrostatics of amino acids and proteins were well regarded by other researchers in the life sciences who applied the tools of physics and chemistry to problems in biology, but the appeal of his work was limited. Even some physical biochemists found his approach disturbingly narrow.[36] In the words of his more clinically oriented colleague, Hastings, "If it hadn't been for World War II, Edwin Cohn might have gone to his grave continuing to measure the dielectric constants of amino acids and proteins with one more decimal of accuracy! The needs of the war literally saved him."[37] Cohn was well situated to participate in building up biomedical research dedicated to the war effort. When asked to contribute his laboratory expertise to the problem of deriving a blood substitute for transfusion from beef blood, he organized his entire department around blood fractionation.

Cohn's singular contribution to the blood substitutes problem was to treat it as a problem in protein chemistry. Ten years of experiments in his laboratory devoted to assessing the effects of solvents, pH, ionic strength, temperature, and protein concentration on solubility of proteins were mobilized for the task of establishing a fractionation method for serum albumin that could be reliably carried out on an industrial scale. The rapidity with which Cohn's group developed a set of methods to accomplish this goal remains striking. They began to work out a method in the late spring of 1940 and settled on the basic framework by the late summer. Rather than relying on salt precipitation, as was conventional for protein purification, their method utilized ethanol at cold temperatures to separate the plasma proteins into five major fractions. This choice enabled straightforward scaling up to industrial scale, for it is much easier to remove alcohol than large amounts of salt (such as ammonium sulfate) from proteins. In particular,

Figure 1. Edwin J. Cohn's own copy of the issue of *Chemical and Engineering News* that celebrated an award from the American Chemical Society for his work on the fractionation of blood plasma. Courtesy Harvard University Archives. Copyright 1948 American Chemical Society. Reproduced with permission.

the recent availability of Sharp and Dohme's freeze-drying machines made it easier to remove the ethanol from the various fractions on a large scale. Besides regulating the ethanol-water ratio and temperature, Cohn's method relied on careful control of three other variables: pH, ionic strength, and protein concentration.[38] Cohn often referred to this approach as the five-variable method, although other researchers and producers quickly termed it the Cohn fractionation method.

Although Cohn was originally asked to develop a method for purifying bovine serum albumin as a possible transfusion material, his suspicion that even a pure beef protein would cause immunological reactions in recipients led him to work on human blood, beginning with plasma purchased from a few professional donors in August of 1940. The same method developed for bovine albumin worked on human albumin, and the promising results led the lab to request human blood from the American Red Cross every week, beginning in February 1941. Cohn also recognized that human plasma proteins in the other fractions had potential medical use. Gamma globulin antibodies could be used for inoculations, isoagglutinins could provide diagnostic tools for blood typing, and blood-clotting proteins could be used to treat disorders such as hemophilia and to prevent extensive bleeding from surgical procedures. The Cohn method was further refined to yield five biologics from human plasma: serum albumin, serum gamma globulin, fibrin foam with thrombin, blood grouping globulins, and fibrin film. In spring 1941 a pilot plant was built at Harvard to scale up production of these materials, starting with serum albumin. Construction was funded by the NRC—the first disbursement of war research funds to Cohn's lab. Early production from the Harvard plant was sufficient to begin clinical trials, and Armour Laboratories in Chicago began to prepare for production of human blood derivatives. When Pearl Harbor was attacked on 7 December 1941, the laboratory was ready to ship human serum albumin to transfuse casualties. In January 1942 the Surgeon General of the Navy authorized contracts for commercial preparation of human serum albumin.[39]

As there was only a small national blood donation service, beginning commercial production of human serum albumin posed an enormous challenge: obtaining enough raw material for the fractionation process. Elliot Robinson, the national director of the Blood Donor Service of the American Red Cross, responded to the sixfold increase in human blood required to produce transfusion materials with dismay:

> The problem of collecting blood in this program is an enormous assembly line, mass production job. . . . The American Red Cross, as you know, is a volunteer organization. . . . The Plasma Program for the Armed Services was . . . an extremely large one. Now in addition . . . we have been asked to collect blood from approximately an additional million and a half donors for the albumin project. . . . Much of the blood is collected in small localities in mobile units. . . . To keep the proper amount of blood going into any given laboratory is quite a task. For example: one of the laboratories

wants 5,000 bleedings per week for plasma and 6,000 for albumin. It is almost an impossible task to get this number of bleedings into one laboratory.[40]

Human albumin and the other blood components were prepared by commercial firms on a large scale by contract with the OSRD-CMR, the Red Cross, and the navy between January 1942 and May 1945. These were no longer simply blood constituents, but were industrially produced therapeutics used for transfusing shock victims, immunizing soldiers, reconstructing the brain covering after neurosurgery, and typing blood.

Rather than handing over direction of the large-scale serum albumin project to the pharmaceutical companies, Cohn became deeply involved in industrial production.[41] The pilot plant for plasma fractionation built in his Harvard laboratory in 1941 allowed his group to develop the industrial-scale production protocols. Besides being the central research and development site, the Harvard plant served as the first large-scale producer for the military. Until commercial production was well under way in 1943, the pilot plant at Harvard operated twenty-four hours a day to provide serum albumin for the armed forces. In addition, Cohn's laboratory trained personnel for the contracted industrial producers: The pharmaceutical houses were required to send representatives for months of training under Cohn to master the procedures.[42]

Cohn also supervised the quality control of the blood-derived therapeutics made by the companies. The navy contracts with Armour Laboratories, Cutter Laboratories, Lederle Laboratories, Eli Lilly and Company, Sharp and Dohme, E. R. Squibb and Sons, and the Upjohn Company stipulated:

The approval and final release of the packaged albumin solution by the Plasma Fractionation Control Laboratory at the Department of Physical Chemistry, Harvard Medical School, Boston, Mass., acting for the U.S. Navy, shall constitute complete fulfillment of the conditions imposed by this paragraph. . . . Under the terms of the contract between the processing laboratory and the Navy, Dr. Cohn is recognized as the consultant in all matters related to the processing method. No variation in the manufacturing process shall be employed without his approval and each laboratory shall be open to inspection by Dr. Cohn or his representative at all times during the usual working hours. In addition, Dr. Cohn will be available for consultation at his office by appointment.[43]

Every lot of blood material sent to the armed forces was checked for purity and safety by members of Cohn's group, and the demarcations between quality control and laboratory research, in terms of both personnel and equipment, were ill defined. The military production contracts allowed Cohn comprehensive authority over the industrial production of the blood-based biologics.

The way Cohn took total control of the Plasma Fractionation Project epitomized his style of leadership. One negative appraisal of Cohn by a Rockefeller Foundation officer after the war reads: "E. J. Cohn is not a first-

Figure 2. Pilot and scaled-up industrial plant for drying human plasma proteins from the frozen state. From Office of Scientific Research and Development, *Advances in Military Medicine* (Boston: Little, Brown, 1948).

class scientist, but is an excellent entrepreneur and promoter. His main weakness however is that he directs his laboratory so completely that the chance of their observing and finding anything 'sideways' is negligible and in this he seems therefore to resemble the worst of industrial research."[44] Cohn's reluctance to relinquish any of the research, development, or quality control work to the processing industries may have suited the military's hierarchical organization, but it proved problematic after the war when laboratories were expected to demobilize back to programs of "pure" research.

Cohn's highly visible contribution to the war effort and his management of the production of therapeutics attest to the existence of a "medical front" that was just as significant a place to wage the war as the distant theaters. Biomedical scientists and the military combined forces to develop laboratory technologies for inoculating and sustaining the fighting bodies abroad. A recognition of the significant role of life science in the design of war medicine is evidenced by Wilfrid Eggleston, whose book *Scientists at War* (1950) provides a patriotic account of Canadian war research:

> The military scientist thrusts fighting man into new frontiers of physical environment, often highly inimical. . . . For these and other reasons, the science of war calls for aid upon the biologist, the physiologist, the bacteriologist, the psychologist, the surgeon, and others, to devise protective and compensatory equipment, so as to make the maximum use out of new discoveries in military fields, to provide a maximum of well-being for the soldier, to modify an unfriendly environment wherever possible; and finally, if in spite of all these measures the fighting man becomes a casualty, to offer whatever may be possible by way of restoration, salvage, and repair.[45]

Cohn's plasma fractionation contributed to the war effort by offering blood-derived medicines to prevent shock and disease as a counterweight to the highly inimical forces of the battlefield. The industrialized killing of twentieth-century war technologies was met on the medical front by industrialized sustenance—a technology of life developed and manufactured in the laboratory of a Harvard biochemist. Authority in military technologies, both for weapons and medicines, was partially displaced from the Pentagon toward the laboratories of prominent university physicists, chemists, and biologists. These topological alterations of the place of the laboratory in the defense of society remained even after the Axis powers were defeated.

Demobilization: Translating National Defense into Public Health

As Michael Dennis has argued, laboratories emerged from World War II as a new source of national strength.[46] Those who had made substantial contributions to war research were celebrated as national heroes, including Edwin Cohn. Articles on Cohn and his contributions to transfusion medicine appeared in *Time* and *Life* magazines, *Collier's*, the *New York Times*, the *Atlantic*,

McCall's, *Ladies' Home Journal*, a television program, a radio program, and international publications.[47] President Truman decorated Cohn with the Medal of Merit. Cohn's achievement had depended on his close association with and direction of industrial producers, as well as on the authority given him by the military. Demobilization called for a new justification of his prominent place in the pharmaceuticals industry, a shift from military medicine to public health. Whereas Cohn and his group regulated the blood processing during the war in the name of national defense, the same activity was now to be pursued for the good of a vulnerable public. Cohn and his laboratory contributed to postwar industrial developments in two ways. First, they continued to operate the pilot plant to improve methods for large-scale production and to provide new therapeutics for clinical testing. Second, Cohn filed patents for his fractionation method, which had been developed before official sponsorship by the OSRD. Exercising his licensing prerogatives allowed him to continue regulating the quality and production of blood-derived medicines. Although Cohn did not receive patent royalties, he required the pharmaceutical companies he licensed to adhere to his own standards for quality control, which were maintained through testing by his laboratory.

In light of the persistent doubts whether it was ethical for medical researchers in universities to take out patents, Cohn had to argue for his proprietary activities with great care.[48] To justify his patents, he presented the pharmaceutical producers, his wartime collaborators, in a new light. Because they would now be marketing the therapeutics for profit, he had to remain involved to protect the public from mercenary interests.[49] The Commissioner of Public Health of Massachusetts asserted that Cohn had a public obligation to oversee production of the blood-derived biologics:

> After reviewing once more the public interest in the further procurement, processing, and distribution of blood derivatives, both in the State of Massachusetts and in the rest of the country, we have come to the conclusion that a continued control of the products that have been developed is essential. Such control is still exercised by the Plasma Fractionation Laboratory at the Harvard Medical School on every product being accepted for the Armed Forces. It would be unjustifiable, therefore, to permit comparable products to be distributed to the public under a less rigid system of control. The purpose behind such control is to assure the public of products of the highest possible quality and safety.[50]

Arguing that he could best protect the public from unsafe products by exercising patents on his methods, Cohn persuaded James Conant, president of Harvard University, to reconsider Harvard's 1935 policy against the patenting of therapeutic discoveries by university researchers. Harvard updated its patent policy in 1950, partly in response to Cohn's ongoing collaborations with pharmaceutical companies. Cohn negotiated patent agreements for collaborative research projects with Armour and Company in

1947 and Sharp and Dohme in 1949. At the same time, Cohn was managing the patents taken out on the original methods developed in the early 1940s. During and immediately after the war, Cohn's plasma fractionation patents were held by the nonprofit Research Corporation in New York, but another organization outside Harvard University was necessary for licensing and regulating the patents.

In December 1944 Cohn established a Commission on Plasma Fractionation and Related Processes, justified on the grounds that government-specified quality control standards for blood products were inadequate. The commission was composed of Cohn; John Enders, a Harvard virologist; Vlado A. Getting, the Commissioner of Public Health for Massachusetts; Roger I. Lee, a member of the Harvard Corporation and president of the American Medical Association; and Dr. George Scatchard, a physical chemist at MIT and a longtime collaborator of Cohn's. The commission contracted with members of Cohn's laboratory to perform the quality control tests, and Cohn's group continued developing new fractionation methods to patent. Consequently, the transition from developing military medicine to licensing patents was negotiated so as to leave the day-to-day control of Cohn's laboratory over plasma fractionation products virtually unchanged.[51]

Establishing a commission to administer a university patent was not unique, but a comparison with the closest precedent sheds light on some novel features of Cohn's postwar commission. In the late 1920s, the Committee on Pernicious Anemia at Harvard Medical School was organized to handle the commercialization of Cohn's method for isolating a liver extract. Harvard had refused to allow Cohn and his collaborator George Minot to patent their method, however, and after two years the Committee on Pernicious Anemia was dissolved, as the project was firmly in industrial hands. It was the short-lived nature of the Committee on Pernicious Anemia that had the longest-range effects for Cohn. Eli Lilly began producing the extract using Cohn's method in 1927, and the company urged Harvard to patent the process, which the university refused to do. In 1928 a Lilly scientist applied for the patent and did not inform Harvard. In addition, from 1927 to 1929, the company marketed the extract and advertised it as "prepared under the direction of the Committee on Pernicious Anemia of the Harvard Medical School."[52] Because Harvard refused to hold the patents, the committee lacked the authority to regulate the uses of Cohn and Minot's method in preparing and marketing the extract. Cohn felt the public would have been better served if he had held the patent and could have regulated the quality of the extract. His insistence on taking out and licensing the patents related to his blood research undoubtedly derived from this earlier experience with Eli Lilly.

In 1950 Cohn began to make plans to regain his patents from the Research Corporation. He founded his own nonprofit patent-holding corporation, the Protein Foundation, charged with "protecting against the improper exploitation of the public where public health and therapeutic procedures are involved." The former president of the Rockefeller Foundation, Chester Barnard, served as the founding chairman of the board. The Protein Foundation was to assimilate the Commission on Plasma Fractionation and expand its mission to include enforcing standards, collecting licensing fees, and increasing public support. As Barnard wrote to his successor at the Rockefeller Foundation: "This blood business is one of the major developments in medical care, in therapeutics and also in prevention. Its importance is indicated by the present somewhat hysterical interest in that fraction known as gamma globulin which contains the viruses that are present in blood and, as I understand it, also the antibodies."[53] Barnard's enthusiasm for the potential of the blood fractionation as a biomedical innovation is striking despite, or because of, his fundamental confusion about whether the gamma globulin fraction contained infectious or curative agents. Besides administering Cohn's patents on plasma fractionation, the Protein Foundation was to "seek funds to promote the manufacture of biomechanical equipment, and to finance other developmental activities."[54]

The degree to which Cohn had successfully interpolated his laboratory into the blood industry is most apparent in the disarray that followed his death. Cohn died rather suddenly in 1953, after several years of dangerously high blood pressure. Hastings, as the chairman of the department of biological chemistry, headed the committee handling Cohn's academic "estate." Hastings referred to the complex set of laboratory functions and adjuncts Cohn had devised as the "Cohn Empire," and complained of the difficulties of disassembling it:

> There was also the problem of the Protein Foundation, and the problem of a Board which has a long and complicated name, which had either to be disbanded or at least dissociated as Harvard's responsibility. Harvard was in big business! There were all sorts of responsibilities that were not academic responsibilities, e.g., being responsible for setting and enforcing standards! It had just developed because it was a good way to let Edwin do this whole thing since it was important for the military aspects of the war, and so it had grown like Topsy and here we were with it![55]

Untangling the university-industry collaboration that had resulted, in Cohn's own words, "on the one hand, in support of fundamental research by Industry and, on the other, in patent agreements with Industry"[56] was complicated indeed. Disbanding his elaborate blood network left only the precedent intact. Even this was out of date: the relationships between universities, industrial producers, and the federal government had been transformed in complicated ways after the war as new science policies were articulated. The direct support of some biomedical research by pharmaceutical companies was largely replaced by federal funding, which was touted as addressing the health concerns of the nation by providing a knowledge base that corporations would use to develop novel procedures and products.

Conclusions

The Plasma Fractionation Project pertains to the more recent ascension of DNA-based biotechnologies in several respects. Alliances between university researchers and pharmaceutical companies are not in fact new: such collaborations were common during the first half of the twentieth century, and their virtual absence from 1950 to 1970 can be traced to abundant (and new) federal funding of biological research. The close ties between Cohn's Harvard laboratory and the pharmaceutical companies in the PFP in particular resemble those linking university molecular biologists and biotechnology companies in the past two decades. In each case "basic" research findings have been commodified and production scaled up, and the university researcher has had an unusually authoritative role in this "application" while remaining a professor. There are also differences between Cohn and the later entrepreneurial molecular biologists, not the least of which is that neither Cohn nor his university profited from his patents.[57]

The postwar expansion of federal funding of basic research in university settings, based on the successes of the scientific mobilization for war, supported a greater division between the "public" and "private" realms of scientific practice. Cohn himself had to contend with the increasingly negative appearance of his collaborations with pharmaceutical companies. He justified his control over protocol development and product quality for blood derivatives in the name of "public interest," finessing the boundary problems. Yet, as was clear after his death, his "empire" was built across the public-private divide. This boundary problem remains acute for those who seek to build biotechnological enterprises out of university laboratories. As biotechnology has boomed during the past decade, however, a renewed faith in the free market's ability to serve the public has opened a new space of tolerance for biology professors who moonlight as entrepreneurs. Moreover, as federal grants have become more difficult to obtain, universities have become more open to pharmaceutical and agribusiness companies as patrons. Although there is concern to maintain legal demarcations between public knowledge (which operates in a tax-free university environment) and private knowledge, the commercialization of molecular biology is reconstituting these social domains.

Two recent events have highlighted the relevance of blood-derived pharmaceuticals to the current ascent of biotechnology. In July 1993 the Food and Drug Administration tightened quality-control standards for collected blood.[58] The community orientation of blood donation services has long protected them from the same scrutiny in regard to safety standards that corporations undergo. Now blood collection centers are being viewed as manufacturing sites and must comply with the same regulatory standards that operate in pharmaceutical plants. Thus the very handling of donors' bodies will now resemble the regulation of laboring microorganisms in biotechnology companies, as each provides material for isolation of biomedicines.

Edwin Cohn's own institution, Harvard Medical School, began in 1993 to build "Institutes of Medicine" to contain both university researchers and commercial ventures, in the hope that the combination will encourage big medical breakthroughs with immediate clinical and commercial application. This ambitious plan draws on Cohn's precedent as Harvard now actively seeks to patent discoveries of therapeutic value, but goes beyond it by housing university researchers' and pharmaceutical companies' laboratories in the same building. The medical school negotiated for funding from the Healthcare Investment Corporation, "the largest venture capital firm in the biotechnology field." This consortium (the planners hope) will tighten considerably the linkages between the clinic, the laboratory, and the bodies of patients, extending the power of the molecular biology laboratory throughout the health care economy and dissolving many of the spatial and social divisions between public and private enterprise. If the *New York Times* is to be believed, Harvard's "Institutes of Medicine," with both university and commercial components, is the way of the future.[59]

World War II spurred the federal government, universities, and industries to collaborate with each other and with the military in unprecedented ways. The success of the mobilization of science for war was used to justify the inauguration of large-scale federal funding of basic research, which enabled an easy differentiation between the public knowledge of universities and the private knowledge of industries. Cooperation between pharmaceutical industries and universities waned until the economic slowdown of the past two decades invited new collaborations—in the name of the national economy, rather than national defense. Biochemists and molecular biologists have used their former independence from practical concerns to claim a special status and access to "fundamental" knowledge of nature that can now be "applied," to salvaging the health of both American citizens and the U.S. economy. And whereas universities such as Harvard once restrained or even prohibited the entrepreneurial activities of the professors, many now expect the patentable discoveries of their faculty to bring in much-needed funding. The current intercalation of academic biologists in biotechnology industries and the prevalence of patenting in university laboratories make Edwin Cohn appear a bit ahead of his time.

Daniel Kevles, Alberto Cambrosio, Arnold Thackray, Jean-Paul Gaudillière, John Edsall, Lily Kay, Evelyn Fox Keller, Barbara Rosenkrantz, Douglas Starr, Jessica Wang, and Leo Slater provided helpful criticisms of earlier drafts of this paper. I am grateful to Richard Wolfe of the Rare Books Room at the Countway Library for assistance in my research in the Cohn papers, and to Frances Coulborn Kohler for deftly editing the paper. Research on this project was supported by postdoctoral fellowships from the National Science Foundation and the Mellon Foundation.

Notes

1. For elaboration see Robert Bud, "The Zymotechnic Roots of Biotechnology," *British Journal for the History of Science* 25 (1992), 127–44. See also Bud, "Biotechnology in the Twentieth Century," *Social Studies of Science* 21 (1991), 415–57; Bud, *The Uses of Life: A History of Biotechnology* (Cambridge/New York: Cambridge University Press, 1993); and the article by Bud in this volume.

2. See, e.g., Edward Yoxen, *The Gene Business: Who Should Control Biotechnology?* (New York: Harper and Row, 1983); Sheldon Krimsky, *Biotechnics and Society: The Rise of Industrial Genetics* (New York: Praeger, 1991); and Michael Fox, *Superpigs and Wondercorn: The Brave New World of Biotechnology and Where It All May Lead* (New York: Lyons and Burford, 1992). All center their historical narratives on the transforming power of recombinant DNA practices.

3. Yoxen, *The Gene Business* (note 2), p. 3.

4. Donna Haraway provides a suggestive historical account of significations of blood and DNA (as bodily fragments) for human relatedness in "Universal Donors in a Vampire Culture: It's All in the Family: Biological Kinship Categories in the Twentieth-Century United States," in *Uncommon Ground: Toward Reinventing Nature,* ed. William Cronon (New York: W. W. Norton, 1995), pp. 321–66.

5. My discussion both here and in the conclusion is informed by a commonsense notion of research in for-profit companies as private and government-supported research in universities as public. As many prominent universities are private, and the majority of large corporations are "publicly" held, the public-private divide is obviously much more complex. (See the article by Martin Kenney in this volume.) Nonetheless, using research knowledge supported by the tax-paying public in a for-profit enterprise is commonly held to raise ethical problems. Daniel Kevles has pointed to similar problems with classified research, as national security considerations impede the paying public's access to research knowledge.

6. Michael A. Dennis, " 'Our First Line of Defense': Two University Laboratories in the Postwar American State," *Isis* 85 (1994), 427–55, on p. 428.

7. Martin Kenney, *Biotechnology: The University-Industrial Complex* (New Haven, Conn.: Yale University Press, 1986), p. 13.

8. Although private philanthropies such as the National Foundation for Infantile Paralysis and the American Cancer Society did support particular areas of basic biochemistry and molecular biology research related to disease, federal funding through the Public Health Service (the NIH) and the NSF supported basic research to an unprecedented degree. See G. Burrough Mider, "The Federal Impact on Biomedical Research," in *Advances in American Medicine: Essays at the Bicentennial,* ed. John Z. Bowers and Elizabeth F. Purcell (New York: Josiah Macy, Jr. Foundation and National Library of Medicine, 1976), vol. 2, pp. 806–71.

9. John P. Swann, *Academic Scientists and the Pharmaceutical Industry: Cooperative Research in Twentieth-Century America* (Baltimore: Johns Hopkins University Press, 1988), p. 1. This article examines biomedical research and the pharmaceutical industries. On the historical relationship between biological research and agriculture, see Charles Rosenberg, "Rationalization and Reality in Shaping American Agricultural Research, 1875–1914," in *The Sciences in the American Context: New Perspectives,* ed. Nathan Reingold (Washington D.C.: Smithsonian Institution Press, 1979), pp. 143–64. Sheldon Krimsky addresses more current issues in the relationship

between agricultural industry and biological research in his contribution to this volume.

10. Charles Weiner, "Patenting and Academic Research: Historical Case Studies," in *Owning Scientific and Technical Information: Value and Ethical Issues*, ed. Vivian Weil and John W. Snapper (New Brunswick, N.J.: Rutgers University Press, 1989), pp. 87–109.

11. Swann, *Academic Scientists* (note 9), p. 170.

12. Vannevar Bush, foreword to James Phinney Baxter III, *Scientists Against Time* (Boston: Little, Brown, 1947), p. vii.

13. Captain Douglas B. Kendrick, Jr. and Lieutenant Commander Lloyd R. Newhouser, "Blood Substitutes in the Military Service," in *War Medicine: A Symposium*, ed. Winfield Scott Pugh (New York: Philosophical Library, 1942), pp. 284–94, on p. 284.

14. Corinne S. Wood, "A Short History of Blood Transfusion," *Transfusion* 7 (1967), 299–303.

15. Charles A. Janeway and J. L. Oncley, "Blood Substitutes," in *Advances in Military Medicine*, ed. Edwin C. Andrus (Boston: Little, Brown, 1947), pp. 444–61, on p. 444.

16. There was a cooperative agreement between the army and the navy that the army would deal with plasma contracts and the navy with plasma fractions. Consequently, all Cohn's interactions were with the navy. Douglas B. Kendrick, *Blood Program in World War II* (Washington, D.C.: Office of the Surgeon General, Department of the Army, 1989, first published 1964), p. 267.

17. See Jonathan Liebenau, *Medical Science and Medical Industry: The Formation of the American Pharmaceutical Industry* (London: Macmillan, 1987).

18. See, e.g., John F. Fulton, "Penicillin, Plasma Fractionation, and the Physician," *Atlantic Monthly* 146 (1945), 107–14.

19. Stephen P. Strickland, *Politics, Science, and Dread Disease: A Short History of United States Medical Research Policy* (Cambridge, Mass.: Harvard University Press, 1972), p. 17, quoting *Hearings on Wartime Health and Education,* 78th Cong., 2nd sess., 2 Jan. 1945, p. 2198.

20. See Lily E. Kay, *Molecules, Cells, and Life: An Annotated Bibliography of Manuscript Sources on Physiology, Biochemistry, and Biophysics, 1900–1960, in the Library of the American Philosophical Society* (Philadelphia: American Philosophical Society, 1989), p. 23.

21. See Kendrick, "Preface," *Blood Program in World War II* (note 16).

22. See Robert E. Kohler, *Partners in Science: Foundations and Natural Scientists, 1900–1945* (Chicago: University of Chicago Press, 1991).

23. See Swann, *Academic Scientists* (note 9).

24. See Daniel J. Kevles, "George Ellery Hale, the First World War, and the Advancement of Science in America," *Isis* 59 (1968), 427–37.

25.Rexmond C. Cochrane, *The National Academy of Sciences: The First Hundred Years, 1863–1963* (Washington D.C.: National Academy of Sciences, 1978). The classic analysis of the policies of the U.S. federal government toward sponsoring science is A. Hunter Dupree, *Science in the Federal Government: A History of Policies and Activities* (Cambridge, Mass.: Belknap Press of Harvard University Press, 1957).

26. Daniel S. Greenberg, "When Science Was an Orphan," in Greenberg, *The Politics of Pure Science* (New York: World Publishing, 1967); and Joel Genuth, "Groping Towards a Science Policy in the United States in the 1930s," *Minerva* 25 (1987), 238–68.

27. The growing literature on the mobilization of scientific research before and during World War II cannot be adequately represented here. The official history of the OSRD is Irvin Stewart, *Organizing Scientific Research for War: The Administrative History of the Office of Scientific Research and Development* (Boston: Little, Brown, 1947). See also A. Hunter Dupree, "The Great Instauration of 1940: The Organization of

Scientific Research for War," in *Twentieth Century Sciences: Studies in the Biography of Ideas*, ed. Gerald Holton (New York: W. W. Norton, 1972), and Nathan Reingold, "Vannevar Bush's New Deal for Research," *Historical Studies in the Physical Sciences* 17 (1987), 299–334.

28. See Daniel J. Kevles, *The Physicists: The History of a Scientific Community in Modern America* (Cambridge, Mass.: Harvard University Press, 1971); Paul Forman, "Behind Quantum Electronics: National Security as a Basis for Physical Research in the United States, 1940–1960," *Historical Studies in the Physical Sciences* 18 (1987), 149–229; and Silvan S. Schweber, "The Mutual Embrace of Science and the Military: ONR and the Growth of Physics in the United States After World War II," Peter L. Galison, "Physics Between War and Peace," and Paul K. Hock, "The Crystallization of a Strategic Alliance: The American Physics Elite and the Military in the 1940s," all three in *Science, Technology and the Military*, ed. Everett Mendelsohn, Merritt Roe Smith, and Peter Weingart (Boston: Kluwer, 1988). For postwar effects of defense patronage see Daniel J. Kevles, "Cold War and Hot Physics: Science, Security, and the American State, 1945–56," *Historical Studies in the Physical Sciences* 20 (1990), 239–64; and Stuart W. Leslie, *The Cold War and American Science: The Military-Industrial-Academic Complex at MIT and Stanford* (New York: Columbia University Press, 1993).

29. Stewart, "Committee on Medical Research of OSRD," *Organizing Scientific Research for War* (note 27).

30. Ibid., pp. 104–15. See also Peter Neuschul, "Science, Government, and the Mass Production of Penicillin," *Journal of the History of Medicine and Allied Sciences* 48 (1993), 371–95; and Harry M. Marks, "Cortisone, 1949: A Year in the Political Life of a Drug," *Bulletin of the History of Medicine* 66 (1992), 419–39.

31. Lily E. Kay, "Selling Pure Science in Wartime: The Biochemical Genetics of G. W. Beadle," *Journal of the History of Biology* 22 (1989), 73–101.

32. National Research Council, Division of Medical Sciences, *Bulletin on Blood Substitutes* 1 (1940), 1–8, courtesy of Douglas Starr and Sam Gibson.

33. A. Baird Hastings, *Crossing Boundaries: Biological, Disciplinary, Human—A Biochemist Pioneers for Medicine* (Grand Rapids, Mich.: Four Corners Press, 1989), pp. 151, 152. On the composition of the NRC committees see Kendrick, *Blood Program in World War II* (note 16), pp. 73–75.

34. Edwin J. Cohn, *University Laboratory of Physical Chemistry Related to Medicine and Public Health, Harvard University: A Brief History of the Support of the Department of Physical Chemistry, Harvard Medical School, 1920–1950*, pamphlet (Harvard University Printing Office, March 1950), courtesy of John T. Edsall.

35. See Weiner, "Patenting and Academic Research" (note 10).

36. Evaluations of Cohn's laboratory by S. P. L. Sörensen, C. R. Harrington, and F. Gowland Hopkins, Rockefeller Archive Center, Tarrytown, New York, record group RF 1.1, series 200D, box 141, folder 1741. See also Robert E. Kohler, *From Medical Chemistry to Biochemistry: The Making of a Biochemical Discipline* (Cambridge: Cambridge University Press, 1982), p. 315. Linus Pauling was Cohn's most serious detractor.

37. Hastings, *Crossing Boundaries* (note 33), p. 71.

38. E. J. Cohn, J. A. Luetscher, Jr., J. L. Oncley, S. H. Armstrong, Jr., and B. D. Davis, "Preparation and Properties of Serum and Plasma Proteins. III. Size and Charge of Proteins Separating upon Equilibration across Membranes with Ethanol-Water Mixtures of Controlled pH, Ionic Strength, and Temperature," *Journal of the American Chemical Society* 62 (1940), 3396–3400.

39. Edwin J. Cohn, "The History of Plasma Fractionation," in Andrus, *Advances in Military Medicine* (note 15), pp. 364–443, on p. 375. On some of the challenges Cohn faced in large-scale production of serum albumin and on the uses of the plasma

fractions on the battlefield, see Douglas Starr, "Dr. Edwin Cohn, the 'King of Blood,' " *Smithsonian* 25 (1995), 124–38.

40. "National Research Council, Division of Medical Sciences, Acting for the Committee on Medical Research of the Office of Scientific Research and Development, Conferences on the Preparation of Normal Human Serum Albumin, at the Department of Physical Chemistry, Harvard Medical School, Faculty Room, Boston, Massachusetts, June 6, 1942," *Memoranda and Communications on the Preparation of Normal Human Serum Albumin*, Rare Books Room, Countway Library of Medicine, Harvard University.

41. Since he contributed to actual industrial development as well as to research of his blood-derived therapeutics, Cohn differed from the physical scientists who worked to realize proximity fuses and missiles. See Dennis, "First Line of Defense" (note 6).

42. Cohn, "History of Plasma Fractionation" (note 39); and interview of Dr. Douglas Surgenor, 5 Aug. 1993, by Angela Creager. Some of those involved in the war project have explained that the decision that Cohn would directly train the industrial producers served to protect the trade secrets of the process, which was classified. (Douglas Starr, personal communication based on interviews with historical participants.)

43. Cohn, "History of Plasma Fractionation" (note 39), p. 378.

44. Excerpt from diary of Rockefeller Foundation officer William F. Loomis, 15 Sept. 1950, Rockefeller Archive Center, record group RF 1.1, series 200D, box 141, folder 1746.

45. Wilfrid Eggleston, *Scientists at War* (London/New York: Oxford University Press, 1950), p. 187.

46. Dennis, "First Line of Defense" (note 6).

47. Copies of many of these articles are preserved in HUG 4290.2, Harvard University Archives, Pusey Library.

48. See Weiner, "Patenting and Academic Research" (note 10).

49. Edwin J. Cohn to James B. Conant, 14 Nov. 1944, in Edwin J. Cohn, *History of the Development of a Patent Policy Based on Experiences in Connection with Liver Extracts and Blood Derivatives, 1927–1951* (University Laboratory of Physical Chemistry Related to Medicine and Public Health publication) (Harvard University Printing Office, April 1951), p. 14; courtesy of John T. Edsall.

50. "Notes on a Conference Between Dr. Vlado A. Getting, Commissioner of Public Health of the Commonwealth of Massachusetts, and Dr. Edwin J. Cohn, at the Department of Physical Chemistry of the Harvard Medical School, October 28, 1944," ibid., p. 11.

51. Some tests of industrial products were carried out in the Massachusetts Antitoxin and Vaccine Laboratory, which had also worked closely with Cohn's lab during the war.

52. Weiner, "Patenting and Academic Research" (note 10), p. 95. I am indebted to Charles Weiner for pointing me to the information on Cohn's earlier experience with Eli Lilly. On the patent conflicts see Cohn to Odin Robert, 5 Nov. 1927, folder "Odin Roberts re: anemia 1927," and letters in folder "George R. Minot and other re: anemia 1927," Cohn papers, Rare Books and Manuscripts, Countway Library of Medicine, Harvard University.

53. Chester I. Barnard to Dean Rusk, 5 March 1953, Rockefeller Archive Center, record group RF 1.1, series 200D, box 141, folder 1746.

54. Appendix E, "Relation of the University Laboratory to the Commission on Plasma Fractionation and Related Processes and Protein Foundation, Incorporated," *Activities of the University Laboratory of Physical Chemistry Related to Medicine and*

Public Health and the Department of Biophysical Chemistry at Harvard University, Rare Books and Manuscripts, Countway Library, E69.10.B2 (1953).

55. Hastings, *Crossing Boundaries* (note 33), p. 153.

56. Cohn, *History of a Patent Policy* (note 49), p. 20.

57. However, Cohn's laboratory did receive at least one research grant of $20,000 from Armour, a pharmaceutical company licensed to use his methods, as part of the collaborative research project between Armour and Cohn's laboratory mentioned above. Conquest to Cohn, n.d., folder "Harvard University 1946–47," Cohn papers, Rare Books and Manuscripts, Countway Library.

58. Philip J. Hilts, "F.D.A. Seeks Tighter Rules on Safety of Blood Supply," *New York Times*, 2 July 1993, p. A11.

59. Susan Diesenhouse, "Harvard's New Test Tube Business," *New York Times*, 22 Aug. 1993, p. D4. Although Diesenhouse does not explore the decision making processes that led to this plan, one professor informed me that there was visible disagreement between the Harvard administration and the medical school over the wisdom of the venture.

Part II
Macromolecular Politics

Diamond v. Chakrabarty and Beyond: The Political Economy of Patenting Life

Daniel J. Kevles

A Matter of Bugs

In 1972 Ananda Chakrabarty, a biochemist at the General Electric Company who had bioengineered a bacterium to consume oil slicks, filed for a patent on the living, altered bacterium. What was patentable according to statute dated back to the patent law of 1793, which declared, in language written by Thomas Jefferson, that patents could be obtained for "any new and useful art, machine, manufacture, or composition of matter, or any new or useful improvement thereof." Jefferson's phrasing remained—and remains—at the core of the U.S. patent code, except for the eighteenth-century word "art," which was replaced in a 1952 congressional overhaul of patent law by the word "process."[1] Chakrabarty contended that his bacterium was a product of his devising, a new composition of matter, and as such patentable under the United States Code governing the subject.

Chakrabarty's application, however, ran up against a longstanding tenet of common patent law that dated back at least to 1889, when, in a landmark ruling, the U.S. Commissioner of Patents rejected an application for a patent to cover a fiber identified in the needles of a pine tree, noting that ascertaining the composition of the trees in the forest was "not a patentable invention, recognized by statute, any more than to find a new gem or jewel in the earth would entitle the discoverer to patent all gems which should be subsequently found." The commissioner added that it would be "unreasonable and impossible" to allow patents on the trees of the forest and the plants of the earth.[2]

The commissioner's ruling formed the basis for what came to be known as the "product-of-nature" doctrine: that while processes devised to extract what is found in nature can be patented, objects discovered there cannot.

They are not inventions, nor can they as a class be made anyone's exclusive property. In the Plant Patent Act of 1930, Congress had granted patentability to one class of living products: plants that could be reproduced asexually. There had been no other extension of patent law to vital entities. Relying on precedent, the United States Patent and Trademark Office denied Chakrabarty's application, citing the product-of-nature doctrine and insisting that any exceptions to it must, like the Plant Patent Act, be authorized by Congress. Chakrabarty appealed his case through the courts, and at the end of 1979 it reached the United States Supreme Court under the rubric of *Diamond v. Chakrabarty*, since the position of the patent office was formally defended by Sidney Diamond, the current patent commissioner.

Patent law is esoteric and highly technical, and it has usually been treated as such by legal historians.[3] However, like many other areas of law, its evolution has tended to be affected by the forces of industrial development and society's reaction to them. Scholars recognize that the commercialization of molecular biology was significantly driven by political and economic considerations, as the contributions by Herbert Gottweis, Martin Kenney, and Susan Wright in this volume attest. Since patents, which have proved essential to the development of biotechnology, establish intellectual property rights, it should not be surprising that political economy has shaped their definition and scope.[4] The Chakrabarty case was no exception. For several years it evolved without fanfare, but by the time it reached the Supreme Court, it was charged with the social and economic stakes that surrounded the swiftly accelerating commercialization of molecular biology.

The scientific key to commercialization was the technique of recombinant DNA coinvented in 1973 by Herbert Boyer and Stanley Cohen, biologists at, respectively, the University of California at San Francisco Medical School and Stanford University. In 1976 Boyer and a venture capitalist named Robert A. Swanson formed the biotechnology firm Genentech—short for "genetic engineering technology"—and in 1978 the company announced in a press conference that its scientists had succeeded in producing human insulin with the aid of recombinant techniques and bacteria. The achievement was heralded in every major newspaper and magazine in the United States except the *New York Times*, which was on strike. *Newsweek* typically proclaimed that "recombinant DNA technology can undoubtedly be used to make scores of other vital proteins, such as growth and thyroid hormones, as well as antibodies against specific diseases."[5]

That expectation stimulated the creation of a biotechnology industry in the United States. New companies were founded at a high pace, while major pharmaceutical firms as well as several oil and chemical giants plunged into recombinant DNA, initiating research programs of their own, letting research contracts to the start-up firms, and even obtaining an equity interest in some of them. Biotechnology firms and firms eager to get into biotechnology sought connections with campuses.[6] In return, the campuses could expect dividends from the biotechnology industry in the form of gifts, research grants, and license fees for the use of patents covering the valuable research products of their laboratories.

Harvard University pointed to the possibilities. In the 1920s Harvard declined in practice to profit from faculty discoveries in public health or therapeutic agents, a practice it transformed into formal university policy in 1934 and 1935, stipulating that neither the university nor its faculty would seek patents in those subject areas "except for dedication to the public." During and after World War II, largely in response to the achievements in blood fractionation accomplished in its laboratories, the university modified its policy, authorizing in exceptional circumstances the patenting and licensing of medical technologies if a public interest would thereby be served. Even so, under the modified policy neither the university nor its employees were permitted to profit from the patents.[7] In 1974, however, the university obtained a package of grants and endowment from the Monsanto Chemical Company that was to total $23 million over a period of twelve years. Harvard applied the grants and income from the endowment to research in a special branch of tumor biochemistry, ensuring its faculty their customary academic freedom. Monsanto would receive an exclusive worldwide license to patents in any inventions or discoveries that might arise in the course of work supported by the company.[8]

In 1975, to bring policy into conformity with practice, Harvard formally adopted a new patent policy that abandoned its earlier commitment to dedicate patents in medical therapeutics and public health to the public. The policy explicitly gave the university the sole right to determine whether to seek a patent covering an innovation in those subject areas, allowed for the sharing of royalties among the inventor, the university, and outside agents, and stipulated that these rules could be overridden by the terms that individual agreements might specify for the disposition of patents arising from the sponsored research. And in 1977 the university established a patent office to encourage the faculty to cooperate in putting the new policy into effect. According to an administrator in the new office, the 1975 policy was "a major change of direction for the university," intended to put it actively in the business of transferring technology into the marketplace and obtaining new resources for research by partnership arrangements with industrial firms.[9]

Most biomedical research in academia was sponsored by grants from the National Institutes of Health (NIH), but NIH had institutional patent agreements with sixty-five universities—Stanford and the University of California among them—allowing them first option on ownership of any inventions that might arise in the course of research supported by the agency. Donald Fredrickson, the head of NIH, noted to a congressional subcommittee in 1976 that the primary aim of the policy was "to facilitate the transfer of technology from the bench to the marketplace by inducing industrial investment."[10] In conformity with that policy, Stanford and the University of California applied for a patent to cover commercial uses of the Cohen-Boyer gene-splicing technique. The charge for a license on that patent was $10,000 at sign-up, $10,000 for each successive year, and a royalty of up to 1 percent on product sales.[11]

Chakrabarty had not used the technique of recombinant DNA to engineer his oil-eating bacterium, but the issue his case raised—the patentability of living organisms—spoke directly to the rapidly increasing stake in biotechnology patents. Ten *amicus curiae* briefs were filed in the case. Most supported Chakrabarty and came from economically interested organizations, including Genentech, the Pharmaceutical Manufacturers Association, the American Patent Law Association, the New York Patent Law Association, and the American Society for Microbiology.[12] The University of California also submitted an *amicus* brief: it was no more alive than other universities to the hopes of revenues from biotechnology, only more immediately interested, by virtue of the activities on its San Francisco campus. Its particular stake in the patenting of living products was echoed and generalized in a single *amicus* brief filed on behalf of the American Society of Biological Chemists, the Association of American Medical Colleges, the California Institute of Technology, and the American Council on Education, as well as several faculty members in biochemistry and molecular biology from Caltech and the University of California at Los Angeles. The brief was unabashedly frank in declaring the fundamental interest of each of these friends of the court in the outcome of the case:

> Some of the *Amici* receive contract funds from commercial corporations whose future funding of research in this field is certain to be influenced by this Court's decision. All of the individual *Amici* receive or plan to receive indirect funding from royalties on patents which are held by their respective universities. The Court's decision in this case will inevitably have a substantial impact on the financing of research at academic institutions, on the way in which research in the laboratories of these *Amici* is financed, and on the probability that research of these *Amici* will be commercially developed so as to find useful, lifesaving and life-improving application. . . . They fear that adoption of a *per se* rule excluding all living things from patentability will inhibit commercial development of the advances they are making in recombinant DNA research.[13]

The only *amicus* brief to side with Diamond was filed first and came from the People's Business Commission (PBC), a public interest group headed by the social activist Jeremy Rifkin. Its objections to Chakrabarty's claim were fundamentally social, economic, and ethical. It first attacked genetic engineering as such, which it warned might "irreversibly pollute the planetary gene pool in radical new ways," and then insisted that the patenting of living products would not be in the public interest. In the PBC's view, patenting life would accelerate the trend in world agriculture toward diminishing diversity in crop genetics and increasing multinational control of the world's food supply. The PBC also insisted that granting Chakrabarty his patent would extend the trend to animals, since allowing patents on microorganisms as new compositions of matter would leave no scientific or legal basis to preclude patenting higher life forms, including mammals and the human manufactures of some Brave New World. Yet what most distressed the PBC—what it saw as "*the essence of the matter*" in the Chakrabarty case—was that to permit patents on life was to imply that "life has no 'vital' or sacred property," that it was only "an arrangement of chemicals, or mere 'compositions of matter.' "[14]

To the other *amici*, especially the two with the greatest immediate interest in the outcome of the case—Genentech and the Pharmaceutical Manufacturers Association—the positions of the government and the PBC seemed to flout fact and logic. Their briefs sought to set matters straight and to provide what amounted to basic instruction in the fundamentals of the patent system, at least as they saw it. Patents did not foster but actually penetrated industrial secrecy to an extent, because they compelled publication of the means and methods that led to a patentable product; denying patents on life would throw corporate recombinant research deeper into the realm of trade secrets and away from public scrutiny, which would be unwise in the socially charged area of genetic engineering. Patents encouraged technological innovation. They had, in fact, done so in the plants area, and they should be allowed to encourage it in genetic engineering, since the field was recognized as a richly promising contributor to the nation's high-technology competitiveness. Speaking from its own experience, Genentech called the patent system at its best "a pro-competitive system," one that could facilitate "the interposition of small but fruitful companies" in industries traditionally dominated by major firms.[15]

The American Patent Law Association took the trouble to point out what should have been obvious to anyone—that living entities (innumerable varieties of domesticated plants and animals) had been treated as property since the advent of acquisitive man. Several of the *amici* conceded that allowing a patent property right in Chakrabarty's bugs might raise the question of the patentability of higher life forms. However, higher life forms were not at issue in the case, only microorganisms. The courts could only resolve the scope of the patentability of life if and when that question came concretely before them, not prospectively. The Pharmaceutical Manufacturers opined that, should the matter arise, it would be easy to draw a line between higher life forms and "the mindless soulless microorganism involved in *Chakrabarty*"; Genentech scoffed at the idea that granting Chakrabarty's claim would permit patents on human beings, declaring that such argument "extends literalism beyond reason."[16]

Genentech supplied a trenchant counter to what the Pharmaceutical Manufacturers termed the "sky-is-falling" issues that had been insinuated into the case. The company's brief contended that it would defeat the patent system to permit controversy to be a criterion of patentability, that the best science and invention were revolutionary and often controversial, and that it was not the province of the Court "to attempt, like King Canute, to command the tide of technological development." The brief added: "The Patent System is, out of necessity, neutral. It cannot be too finely tuned to the kind (as distinguished from the quality) of creation involved. . . . Most particularly must it abjure prior restraints, because they chill expression in literature and science alike. The neutrality of the Patent and Trademark Office requires that it leave to other agencies the regulation of technology, after the fact of its creation."[17]

On 16 June 1980 the United States Supreme Court held by a vote of five to four that Chakrabarty had a right, within existing statutes, to a patent on his microorganism. Chief Justice Warren Burger delivered the majority opinion, praising the broad language that Thomas Jefferson had written into the patent law of 1793, which remained at the core of the patent code: he called it expressive of its author's "philosophy that 'ingenuity should receive a liberal encouragement' " and noted that all succeeding Congresses had left Jefferson's language virtually intact. Rejecting the contentions of the Patent Office, he found that the patent code as written was ample enough to accommodate inventions in areas unforeseen by Congress, including genetic technology, and to cover living microorganisms. Congress, in passing the Plant Patent Act of 1930, had "recognized that the relevant distinction was not between living and inanimate things, but between products of nature, whether living or not, and human-made inventions." Chakrabarty's bugs were new compositions of matter, the product of his ingenuity, not of nature's. As such, they were patentable under existing law. The minority's opinion, delivered by Justice William Brennan, argued precisely the opposite—that, in view of the legislative history of the two plant acts, the extension of patent protection to living microorganisms required new law.[18]

Both the majority and the minority agreed that the question before the Court was the narrow one of statutory interpretation. Justice Lewis F. Powell, Jr., however, dissented in the case, as he wrote to Brennan, because of "the relative novelty of patenting a living organism, and by my conviction that the issue should be decided by Congress." At Powell's urging, Brennan's opinion included the observation—Powell wrote the passage—that the case concerned a composition that "uniquely implicates matters of public concern" and advanced that fact as a special reason for Congressional jurisdiction. Chief Justice Burger also addressed the apprehensions of the Patent Office and the PBC concerning the "grave risks" in genetic engineering, writing in his opinion that their briefs "present a gruesome parade of horribles" and reminding the Court "that, at times, human ingenuity seems unable to control fully the forces it creates." Burger observed, however, that genetic research with its attendant risks would probably proceed with or without patent protection for its products, and that neither legislative nor judicial fiat as to patentability would "deter the scientific mind from probing into the unknown any more than Canute could command the tides." More important, the Court was "without competence" either to brush aside the horribles "as fantasies generated by fear of the unknown, or to act on them." Matters of high policy, embodying competing interests and values, were best handled by Congress and the Executive Branch—the political rather than the judicial branches of the government. The Court's task was the "narrow one of determining what Congress meant by the words it used in the statute"—which the Court had done—and once that was accomplished, its powers were "exhausted."[19]

Congress appeared uninterested in restricting the scope of what could be patented. Quite the contrary: at the end of the 1970s the commercial prospects of biotechnology figured prominently in the minds of federal policy makers. Since early in the decade, the U. S. annual trade balance in manufactured goods that depended little on research had grown more negative, doubling to a deficit of some $35 billion. The loss had been offset by a trade surplus in high-technology goods that had almost quadrupled, reaching about $39 billion in 1979. There was a broadening consensus that "*innovation has become the preferred currency of foreign affairs*," as Howard Bremer, the patent counsel of the Wisconsin Alumni Research Foundation, advised a House committee. Yet the relative power of American innovation seemed to be falling. The number of patents issued by the United States in 1978 was roughly the same—about 66,000—as it had been in 1966; the proportion issued to foreigners had nearly doubled, from 20 percent to almost 40 percent. Among the leading high-tech products of the 1970s were computers, semiconductors, chemicals, and pharmaceuticals; molecular biology was widely regarded as a leading prospect for the 1980s.[20] In 1980 Congress explicitly encouraged universities to patent and privatize the results of federally sponsored high-technology research.

Oyster Rights

After the *Chakrabarty* ruling, several commentators echoed the PBC's insistence that despite the caveats of the lawyers and the courts, the decision appeared to leave no legal obstacle to the patenting of higher forms of life—or, by implication, to the genetic engineering of such life forms.[21] The legal opening ultimately came to the attention of three marine biologists in Washington State—Standish K. Allen, Jr. and Sandra L. Downing of the University of Washington and Jonathan A. Chaiton of the Coast Oyster Company. In September 1984 they applied for a patent on an improved version of *Crassostrea gigas*, a variety of the Pacific oyster. The claim was partly for a process that made the oyster more edible. However, it also covered the improved oyster as such, which challenged precedent.[22]

Standish Allen had first learned to improve oysters—an Atlantic variety—

as a junior member of a research team at the aquaculture station of the University of Maine, where he began his graduate training in 1976. The improvement was accomplished by treating fertilized brood stock with a chemical called cytochalasin B so that the oysters developed as triploids—that is, creatures with three sets of chromosomes instead of the normal two. Allen, who is now a staff member of the Rutgers Shellfish Research Laboratory in Port Norris, New Jersey, has explained that "triploidy makes oysters sterile, which means that they don't put energy into reproduction," adding, "This gives them more energy to devote to their overall bodily growth and quality." In fact, the triploid Atlantic oysters weighed 40 percent more at maturity than their diploid counterparts. Allen remembers, "I then had the great idea of trying to sell commercial hatcheries on triploids. But oyster farming in Maine was just a fledgling industry; there were few hatcheries and they were struggling to stay alive. Maine oystermen weren't interested in technology. I felt shunned, so I went out to the University of Washington Department of Fisheries to get a Ph.D. with research on fish."[23]

Allen stuck to fish until the summer of 1983, when Sandra Downing, who was then a master's degree candidate in the Fisheries Department, drew him back into oyster research. She had learned about his work on Atlantic oysters and been stimulated to begin a project on triploidy in Pacific oysters. Downing proposed to obtain the triploids by subjecting newly fertilized oyster eggs to between six thousand and ten thousand pounds of hydrostatic pressure. Hydrostatic treatment had never been used with oysters, but it had been successfully employed at the University of Washington to make triploid salmon. In the late fall of 1983 Downing discussed the triploid project at a meeting on oyster broodstock held in Astoria, Oregon, and members of the Pacific Coast Oyster Growers Association encouraged her to press ahead with the work.[24]

Oyster farming on the north Pacific Coast, unlike its counterpart on the Atlantic, was a thriving, hatchery-oriented industry. It was dominated by the Coast Oyster Company, which was owned by Hilton Seafoods. Coast Oyster operated some 22,000 acres of farms spread along the tidelands of Washington, Oregon, and California, produced more than twenty billion oyster larvae a year, and was one of the largest oyster enterprises in the world. The company was headed by Vern Hayes, a man in his seventies who had spent much of his life in the oyster business and who saw definite economic potential in triploidy. He hired Jonathan Chaiton, who had recently graduated from the University of Washington, to do research on triploids, and he made Chaiton, as well as Coast's hatchery facilities at Quilcene, Washington, available to Allen and Downing. Allen recalls, "Hayes was himself an immense enthusiast of the project. He always wore a blue pin-stripe suit. He'd come down to the beds, take off his jacket, put on galoshes. He'd work with me in the muck with his tie still on."[25]

Species experts know that the Pacific oyster is distinguishable from its

Atlantic cousin by its gray and purple coloring and by its larger adult size. The work in the muck revealed that it is also distinguishable in how it responds to having three chromosomes. The Pacific oyster is normally fit for the human palate only nine months out of twelve. During the other three, it diverts so much of its energy to reproductive duty as to make its meat soft and sour. Allen and his collaborators found that the induction of triploidy did not add weight to the mature Pacific oyster, as it did with the creature's mature Atlantic cousin. Instead, the sterility it fostered made the oyster edible—and hence marketable—all year round.[26]

That outcome was unexpected, but it was evident enough by the spring of 1984 to prompt Coast Oyster to consider filing a patent claim on the production of triploid-sterile oysters by the hydrostatic method. Downing remarks that she was "just interested in the science, not the commercial aspects of the work," and Allen says that the idea for submitting the patent claim was not his, adding, "It originated with Vern Hayes. I was naive about things like patents in those days." Hayes was not naive. Hilton Seafoods reached an agreement in principle with the University of Washington. The company would pay the costs of filing and pursuing the patent, full ownership of which Allen, Downing, and Chaiton would assign to the university. In exchange, the university would give Coast Oyster an exclusive license to the method, one good throughout the world except in the operating areas of the Pacific Coast Oyster Growers Association.[27]

The patent application was drawn up by David Maki, a lawyer with Seed and Berry, a Seattle firm specializing in intellectual property law that had represented Coast Oyster for a number of years. Maki explains why he extended the claim beyond the hydrostatic process to incorporate the triploid oyster—the living product itself: "I wanted to provide maximum protection for my client. Besides, there was a transition occurring in case law on living organisms, and there were rumors around that the scope of patentability might be enlarged to include living animals."[28]

The transition had not yet reoriented the examiners in the U.S. Patent Office. They denied the claim, holding that neither *Diamond v. Chakrabarty* nor any other patent ruling authorized the grant of a patent on a higher animal, even a mere invertebrate. The examiners also found that the triploid oyster was not patentable on the technical ground that the innovation was obvious to anyone schooled in the art of oyster breeding. Allen and his colleagues appealed the examiners' decision to the Board of Patent Appeals and Interferences of the U.S. Patent and Trademark Office. The Board of Appeals could have pointed to the limited scope of *Diamond v. Chakrabarty* and found that Congressional action was necessary to extend patent protection further to living organisms. It had already cast a vote against Congress and for biotechnology in 1985, however, when it reviewed the patent application of Kenneth Hibberd, a scientist at a subsidiary of Molecular Genetics Research in Minnetonka, Minnesota.

Hibberd had applied for a patent under the industrial patent laws for a type of genetically engineered corn—"a maize seed having an endogenous free tryptophan content of at least one-tenth milligram per gram dry seed weight and capable of germinating into a plant capable of producing seed" with the same level of free tryptophan. Although the examiners had acknowledged that the innovation fell within the scope of *Chakrabarty*, they had denied Hibberd's application, claiming that Congress had intended plants to be protected exclusively under the Plant Patent Act and the Plant Variety Protection Act (PVPA) of 1970. However, the Patent and Trademark Appeals Board awarded Hibberd his patent, holding, in *Ex parte Hibberd*, that the basic utility patent law "has not been narrowed or restricted" by the passage of the Plant Patent Act or PVPA, that it predated both acts, and that—with genuflection to *Diamond v. Chakrabarty*—these plant-specific acts did not "represent exclusive forms of protection for plant life."[29] In 1987, in the oyster case, the Board cast another vote for legal logic and, in consequence, for biotechnology, issuing the decision known since as *Ex parte Allen*. It upheld the examiners on the point that obviousness of art disqualified the oyster for a patent, but it also declared that patents could in principle be granted on nonhuman animals.[30]

Animals by Design

Ex parte Allen meant a good deal to Harvard University and Philip Leder. A distinguished biomedical scientist, Leder was appointed to the faculty of the Harvard University Medical School in 1981. In conjunction with his recruitment, the DuPont Corporation had that year given Harvard $6 million for support of Leder's research. A Harvard spokesman stressed that the grant was "sympathetic with academic freedom and the pursuit of knowledge," declaring that it imposed no restraints on the recipients' freedom to talk or publish.[31] The principal quid pro quo was simple. While Harvard would own any patents that might arise from Leder's research, DuPont would be entitled to an exclusive license on any and all such properties.

During the next two years, Leder and his collaborator Tim Stewart developed a so-called oncomouse—a mouse genetically engineered to be supersusceptible to cancer. They accomplished the feat by exploiting the then-recently-developed transgenic technology to insert the *myc* oncogene, tied to a mammary-specific promoter, into the new embryo of a normal mouse. The work was not done for the sake of devising a patentable product, but once it was accomplished, Leder recognized that it might have commercial possibilities. About the end of 1983 he brought his mice to the attention of the Office of Technology Licensing and Industry Sponsored Research, the recently established patents arm of the Harvard Medical School.[32]

To explore the issue, the Office of Technology Licensing assembled a small group, including, along with Leder and several DuPont intellectual property lawyers, Paul Clark, a patent attorney from the downtown Boston law firm of Fish and Richardson, Harvard's principal outside patent counsel. Clark later recalled that "the work's most apparent and compelling manifestation was the animal itself," continuing, "it became clear immediately that it was important to claim the mice, to give Harvard and its licensee, DuPont, all the legal rights to which they were entitled. Claims on methods of using the mice, or on plasmids, although of some importance, would not have adequately protected the invention."[33] Clark's reasoning was standard among patent lawyers: better to protect the product as well as the processes used to produce it; otherwise competitors would use different processes to develop similar products.

Clark also saw that Leder's transgenic animals were, like the bacteria in *Chakrabarty*, new compositions of matter made by humans, and he knew that the Supreme Court had admonished in the *Chakrabarty* case that a court cannot properly consider the state of being alive when deciding whether something falls within the protection of patent law. Thus, Clark explains, "it was hard for me to see any legal basis for excluding claims on animals."[34]

On 22 June 1984 Clark filed an application for a patent on behalf of Harvard University on Leder and Stewart's invention. The main utilities that he claimed were straightforward, including the use of such animals as sources of malignant or proto-malignant tissue for cell culture and as living systems on which to test compounds for carcinogenicity or, in the case of substances like vitamin E, power to prevent cancers. However, Clark was not at all conservative in what he claimed as the actual invention. It was not simply a transgenic mouse with an activated *myc* gene, which would have been extraordinary enough. It was any transgenic mammal, excluding human beings, containing in all its cells an activated oncogene that had been introduced into it—or an ancestor—at an embryonic stage. Following *Ex parte Allen*, the patent examiners had no problem granting Leder and Stewart's claim. And in April 1988 a U.S. patent was awarded to Harvard University on any nonhuman mammal transgenically engineered to incorporate in its genome an oncogene tied to a specific promoter.[35]

The *Ex parte Allen* ruling had provoked outcries from Jeremy Rifkin and his followers, who invoked essentially the same arguments against the patenting of life as the PBC had in its *Chakrabarty* brief. The patenting of animals, however, brought new groups into the debate—notably animal rights activists, environmentalists, clerics, and farmers' representatives. Their objections were well aired in 1987 and 1989 in hearings held before the House Judiciary Subcommittee that dealt with patents, which was chaired by Rep. Robert Kastenmeier. The objections raised to patenting animals tended to be specific to the groups raising them: animal rights activists contending that such patents would exacerbate the degradation of animals; environmentalists arguing that genetically engineered animals would escape and threaten the integrity of wildlife; clerics claiming that patenting reduced God's creatures to mere material objects; and farm spokespersons worrying about the economic effects of patented animals on small farmers.[36]

Strong defenses of animal patenting came from other witnesses, notably representatives of the biotechnology industry and of major universities. Their arguments, echoing those advanced in the large majority of the *amicus* briefs submitted in the *Chakrabarty* case, emphasized the role of patents in stimulating biotechnological innovation, fostering American competitiveness, and advancing medical research, including diagnostics, therapies, and cures. No significant objection was raised against animal patenting by university representatives or scientists on grounds that such patenting would impede access to or use of transgenic research materials.

Kastenmaier had called the hearings partly in response to the outcry but also in an attempt to assert congressional authority over policy for patenting living organisms. He thought that the extension of patent protection to higher organisms constituted a "quantum leap," one that the Patent Office had been high-handed in making. In his view, the Constitution gives power in determining the scope of patents to the Congress.[37] Kastenmeier's district was centered on Madison, Wisconsin, and included the University of Wisconsin community and its biotechnological spin-offs as well as a considerable agricultural constituency. He and his subcommittee responded to the debate pragmatically—ignoring most of the objections raised by Rifkin and his allies but paying attention to those touching directly on issues of public policy that concerned the key interest groups involved, particularly agriculture. In 1988 Kastenmeier produced a bill exempting farmers from any restraint, including the restraint of royalty payments, on what they did with the progeny of their patented animals. It declared explicitly that human beings cannot be patented. The bill passed the House, but it was not taken up in the Senate before the end of Congress.[38] Since then, no bill addressing animal patents has reached the floor of the House or Senate. The reason has been not so much the force of the legal logic advanced by the U.S. Patent Office as the force of the biotechnology complex in the political economy of intellectual property.

Notes

1. *Diamond v. Chakrabarty*, 447 U.S. (*United States Supreme Court Reports*) 303, 100 S. Ct. (*Supreme Court Reporter*) 2204 (1980), p. 2207.

2. *Ex parte Latimer*, 12 March 1889 (C.D., 46 O.G. 1638), U.S. Patent Office, *Decisions of the Commissioner of Patents and of the United States Courts in Patent Cases . . . 1889* (Washington, D.C.: U.S. Government Printing Office, 1890), pp. 123–27. See also H. Thorne, "Relation of Patent Law to Natural Products," *Journal of the Patent Office Society* 6 (1923), 23–28.

3. An exception to the rule in the legal history of patents is the indispensable Stephen A. Bent et al., *Intellectual Property Rights in Biotechnology Worldwide* (New York: Stockton Press, 1987). Also helpful is Friedrich-Karl Beier, R. S. Crespi, and Josef

Straus, *Biotechnology and Patent Protection: An International Review* (Paris: Organization for Economic Cooperation and Development, 1985).

4. Several recent works have pioneered the treatment of intellectual property in terms of political economy. The key work for plants is Jack R. Kloppenburg, Jr., *First the Seed: The Political Economy of Plant Biotechnology, 1492–2000* (New York: Cambridge University Press, 1988), which can be supplemented by Calestous Juma, *The Gene Hunters: Biotechnology and the Scramble for Seeds* (Princeton, N.J.: Princeton University Press, 1989); and Glenn Bugos and Daniel J. Kevles, "Plants as Intellectual Property: American Law, Policy, and Practice in World Context," *Osiris* 7 (1992), 119–48. The literature on intellectual property in biotechnology is vast. Several key works provide excellent entry: Charles Weiner, "Professors, and Patents: A Continuing Controversy," *Technology Review* 89 (Feb.–March 1986), 33–43; Weiner, "Patenting and Academic Research: Historical Case Studies," *Science, Technology, and Human Values* 12 (1987), 50–56; William H. Lesser, ed., *Animal Patents: The Legal, Economic, and Social Issues* (New York: Stockton Press, 1989); and Sheldon Krimsky, *Biotechnics and Society: The Rise of Industrial Genetics* (New York: Praeger, 1991).

5. Steven S. Hall, *Invisible Frontiers: The Race to Synthesize a Human Gene* (New York: Atlantic Monthly Press, 1987), pp. 87–88, 199–203, 213–22, 231–35, 241–48, 266, 269–83; and Matt Clark with Joseph Contreras, "Making Insulin," *Newsweek*, 18 Sept. 1978, p. 93 (quotation).

6. Susan Wright, "Recombinant DNA Technology and Its Social Transformation, 1972–1982," *Osiris* 2 (1986), 303–60. See also Martin Kenney, *Biotechnology: The University-Industrial Complex* (New Haven, Conn.: Yale University Press, 1986), pp. 44–45, 56, 61–67, 73, 78–80, 140, 191; Nicholas Wade, "Recombinant DNA: Warming Up for the Big Payoff," *Science* 206 (1979), 663, 665; and Marilyn Chase, "Search for 'Superbugs': Industry Sees a Host of New Products Emerging from Its Growing Research in Gene Transplants," *Wall Street Journal*, 10 May 1979, p. 48.

7. Harvard University, *A Statement of Policy in Regard to Patents on Discoveries or Inventions Bearing on Health and Therapeutics, Conforming to Action of the President and Fellows on May 28, 1934, and May 20, 1935* . . . ; Edwin J. Cohn, *History of the Development of a Patent Policy Based on Experiences in Connection with Liver Extracts and Blood Derivatives, 1927–1951* (Cambridge, Mass.; University Laboratory of Physical Chemistry Related to Medicine and Public Health, Harvard University, 1951); and Weiner, "Patenting and Academic Research" (note 4). For details on the evolution of Harvard's policy in relationship to the work of Cohn, see the essay in this volume by Angela Creager, to whom I am indebted for a copy of Cohn's *History*.

8. Kenney, *Biotechnology* (note 6), pp. 58–59.

9. Harvard University, *A Statement of Policy in Regard to Patents and Copyrights as Adopted by the President and Fellows on November 3, 1975*; and interview of Steven Atkinson (Office of Technology Licensing, Harvard Medical School) by Daniel J. Kevles, 19 March 1990.

10. Donald Fredrickson, statement and testimony, Senate Judiciary Committee and Labor and Public Welfare Committee, Joint Hearing, *Oversight Hearing on Implementation of NIH Guidelines Governing Recombinant DNA Research*, 94th. Cong., 2nd Sess., 22 Sept. 1976. See also Nicholas Wade, "Gene Splicing: At Grass-Roots Level a Hundred Flowers Bloom," *Science* 195 (11 Feb. 1977), 558–60, repr. in *The DNA Story: A Documentary History of Gene Cloning*, ed. James D. Watson and John Tooze (San Francisco: W. H. Freeman, 1981).

11. By 1984 the revenues—shared by Stanford and the University of California—exceeded $750,000 annually and were predicted to total between $250 million and $750 million over the life of the patent. Kenney, *Biotechnology* (note 6), pp. 258–59; and *Wall Street Journal*, 3 Aug. 1981.

12. *Wall Street Journal*, 15 Jan. 1980; and Nicholas Wade, "Supreme Court Hears Argument on Patenting Life Forms," *Science* 208 (4 April 1980), 31–32. The *amicus* briefs are with *Diamond v. Chakrabarty* (note 1) [Docket No. 79–136].

13. *Brief Amicus Curiae of the Regents of the University of California*, Jan. 1980; and *Brief of Dr. Leroy Hood, Dr. Thomas Maniatis, Dr. David S. Eisenberg, the American Society of Biological Chemists, the Association of American Medical Colleges, the California Institute of Technology, the American Council on Education as Amicus Curiae*, 28 Jan. 1980.

14. *Brief on Behalf of the People's Business Commission, Amicus Curiae*, 1979. For the deeper cultural roots of the controversy over agricultural biotechnology, which continues, see Sheldon Krimsky's article in this volume.

15. *Brief on Behalf of the Pharmaceutical Manufacturers Association, Amicus Curiae*, 1980; and *Brief on Behalf of Genentech, Inc., Amicus Curiae*, 1980.

16. *Brief on Behalf of the American Patent Law Association, Inc., Amicus Curiae*, 1980; and *Brief on Behalf of the Pharmaceutical Manufacturers Association, Amicus Curiae*, 1980; *Brief on Behalf of Genentech, Inc., Amicus Curiae*, 1980.

17. *Brief on Behalf of the Pharmaceutical Manufacturers Association, Amicus Curiae*, 1980; and *Brief on Behalf of Genentech, Inc., Amicus Curiae*, 1980.

18. *Diamond v. Chakrabarty* (note 1), pp. 2206–12.

19. Ibid.; and Lewis F. Powell to Mr. Justice Brennan, 29 May 1980, William J. Brennan MSS, Library of Congress, container 535.

20. Wright, "Recombinant DNA Technology" (note 6), pp. 320, 337–38; Howard W. Bremer, statement, House Committee on the Judiciary, *Industrial Innovation and Patent and Copyright Law Amendments: Hearings before the Subcommittee on Courts, Civil Liberties, and the Administration of Justice of the Committee on the Judiciary*, 96th Cong., 2nd sess., 3, 15, 17, 22 April, 8 May, 9 June 1980; Donald R. Dunner, statement, ibid.; and Tom Railsback, remarks before the House, *Congressional Record*, 17 Nov. 1980, p. 29897.

21. "Blue Chips for a Biochemist," *Time*, 9 March 1981, p. 57; Peter Gwynne, "Court Decision Spurs Genetics Research," *Industrial Research and Development* 22 (Aug. 1980), 45–46; Debra Whitfield, "Patent Decision Could Spur Genetic Research Industry," *Los Angeles Times*, 17 June 1980, p. 11; and C. Larry O'Rourke, "The Chakrabarty Decision," *Environment* 22 (July–Aug. 1980), 5.

22. Copy of original patent application, 6 Sept. 1984, pp. 12–23 in the *Joint Appendix*, filed 24 Nov. 1987, to U.S. Court of Appeals for the Federal Circuit, Appeal No. 87-1393, *In re Standish K. Allen, Jonathan A. Chaiton, and Sandra L. Downing*.

23. Telephone interview with Standish Allen by Daniel J. Kevles, 29 Sept. 1989. The chemical treatment prevented the fertilized oyster cells from completing the first phase of meiosis, during which they rid themselves of one set of chromosomes. See also Standish K. Allen, Jr., "Triploid Oysters Ensure Year-Round Supply," *Oceanus* 31 (Fall 1988), 60; Jon G. Stanley, Standish K. Allen, Jr., and Herbert Hidu, "Polyploidy Induced in the American Oyster, *Crassostrea virginica*, with Cytochalasin B1," *Aquaculture* 23 (1981), 1–2, 8; and Stanley, Hidu, and Allen, "Growth of American Oysters Increased by Polyploidy Induced by Blocking Meiosis I but Not Meiosis II," *Aquaculture* 37 (1984), 147–48, 153.

24. Telephone interview with Sandra Downing, by Daniel J. Kevles, 16 Oct. 1989. See also *Polyploid Press*, May 1984, pp. 1–2, attached to Ken Cooper (Hilton Seafoods/Coast Oyster Co.), memorandum, "Polyploid Pacific Oyster Patent and Licensing Agreement," 4 Aug. 1987, copy in possession of Daniel J. Kevles; and copy of original patent application, in *Joint Appendix* (note 22), pp. 13–14. Like the chemical treatment, the hydrostatic pressure blocked the egg from expelling the third set of chromosomes and thus made it triploid. Downing and Allen employed cytochalasin B to produce triploid controls, but since chemicals created problems of certification when used with animals destined for human consumption, they initially had no

plans to rely on chemical induction at commercial facilities (*Polyploid Press*, May 1984, p. 2).

25. Allen interview (note 23); telephone interview with Ken Cooper by Daniel J. Kevles, 2 Oct. 1989; and Cooper, memorandum, "Polyploid Pacific Oyster Agreement" (note 24), pp. 2–3.

26. "Reply Brief to Examiner's Second Supplemental Answer and Amendment Pursuant to Rule 116," 5 Jan. 1987, in *Joint Appendix* (note 22), pp. 69–70, 79; and "Brief for Appellant," filed 24 Aug. 1987, *In re Allen, Chaiton, and Downing* (note 22), p. 10.

27. "Brief for Appellant," p. 10; Allen, Downing, and Cooper interviews (notes 23, 24, 25); Cooper, memorandum, "Polyploid Oyster Patent Agreement" (note 24), pp. 2–5, and attached letters: B. F. Berry to Donald R. Baldwin, 22 Aug. 1984, and Baldwin to Berry, 13 Feb. 1985. After the agreement was reached in principle, Coast Oyster provided funds—eventually amounting to $6,300—to the University of Washington to help support the project (Cooper memorandum, pp. 5–6).

28. Telephone interview with David Maki, by Daniel J. Kevles, 3 Oct. 1989.

29. *Ex parte Hibberd et al.* (1985) 227, *United States Patent Quarterly*, 443.

30. *Ex parte Allen* (1987) 2, *United States Patent Quarterly* (2nd ser.), 1425.

31. *Harvard Crimson*, 30 June 1981, pp. 1, 6; *Harvard University Gazette*, 2 July 1981, p. 1.

32. Interview with Philip Leder, by Daniel J. Kevles, 21 June 1988.

33. Paul T. Clark, letter to Daniel J. Kevles, 12 April 1990.

34. Clark to Kevles (note 33).

35. Leder et al., *Transgenic Non-Human Animals, United States Patent Number 4,736,866*, 12 April 1988.

36. House Committee on the Judiciary, *Patents and the Constitution: Transgenic Animals: Hearings before the Subcommittee on Courts, Civil Liberties, and the Administration of Justice*, 100th Cong., 1st sess., 11 June, 22 July, 21 Aug., 5 Nov. 1987; and House Committee on the Judiciary, Subcommittee on Courts, Intellectual Property, and the Administration of Justice, *Transgenic Animal Patent Reform Act of 1989, H.R. 1556*, 101st Cong., 1st sess., 13, 14 Sept. 1989.

37. Interview with Robert Kastenmeier, by Daniel J. Kevles, 10 Oct. 1991.

38. The bill was H.R. 4970, 100th Cong., 2nd sess., *Congressional Record*, 13 Sept. 1988, pp. H7436–38.

Molecular Politics in a Global Economy

Susan Wright

Scientific knowledge in the late twentieth century is an intensely contested resource over which nations, corporations, universities, and scientists struggle at many levels—local, national, international, and transnational. The case considered here—the rise and fall of genetic engineering controls in the United States and the United Kingdom from 1972, when results of the first controlled genetic engineering experiments were published, to 1982, when controls for research and industrial processes in both countries were largely dismantled—provides a window on these struggles. There is a danger that this history will be written off as unimportant, a mere perturbation in the advance of science and technology. In contrast, I argue that it reveals a great deal about the political economy of science, that is, the mix of political and economic interests affecting the disposition of science in a world economy characterized by the increasing globalization of trade, production, and finance. Rarely do those engaged in initiating a new field pause to reflect on its possible consequences. The pause taken with respect to genetic engineering, and the subsequent contests over who should control the field and how, meant that social interests in the field were documented in immense detail, providing an opportunity to examine the fine structure of the social dynamic of scientific development and change.

This essay examines the rise and fall of the American and British controls for genetic engineering in terms of three main phases, characterized by the nature of the policies being developed.[1] The first phase, lasting from 1972 to 1975, was characterized by efforts in both countries to reduce the problems posed by the new field to technical ones. In the second phase, lasting from 1976 to 1978, the two policies diverged in terms of scope, legal force, and, to an extent, problem definition. In the third phase, from 1979 to 1982, the scope and legal standing of the two policies continued to diverge but their content converged, each now embracing similar discourses concerning the problem and each endorsing a dramatic weakening of the original controls.

Each phase involved strong contests in formal arenas as well as in informal negotiations off-stage, and a variety of social groups and institutions engaged the issue at various levels—local, national, international, and transnational. But how should these contests be analyzed? How should relations among groups and institutions be addressed? In responding to such questions, I draw on two main traditions. The first is the voluntarist tradition that runs through liberal political philosophy as well as much of Marx; that tradition assumes that the historical subject has an essential explanatory role, rooted in the ideas of human agency and human responsibility, although it also recognizes that power is generally exercised within structurally determined limits.[2] The second is the poststructuralist tradition associated with the work of Michel Foucault; it addresses the forms of power embodied in particular institutions, discursive practices, forms of knowledge, and in the constitution of human beings as subjects and objects.[3] These traditions, both of which have inspired rich empirical studies, seem to be in conflict. I want to argue that a synthesis of these traditions is not only possible but desirable and productive.

Toward a Synthesis of Postpluralism and Poststructuralism

In the 1960s, a long and productive debate within the voluntarist tradition focused on the question of the location and expression of power. Pluralists like Robert Dahl claimed that power was distributed throughout democratic societies and that its operation could be determined by investigating observable behavior, especially decisions taken in formal policy arenas.[4] In contrast, postpluralists like E. E. Schattschneider, Peter Bachrach, Morton Baratz, Frederick Frey, and Steven Lukes challenged the pluralist assumption that formal policy arenas are level playing fields and assumed generally that power might operate most crucially and influentially outside the formal arena. It was essential, in their view, to address this "second face of power"—that is, the informal processes that may affect the form of the policy arena, the selection of decision makers, the organization of the agenda, and other less visible characteristics of policy processes.[5]

Further debate turned on the question of just how far this analysis could be taken. The problem is nicely illustrated by Peter Saunders in a diagram depicting how various possible issues are shaped in one way or another (Figure 1): level 1, consideration of issues that make it into the formal arena; level 2, actions that decide whether issues will be addressed or not; level 3, actions that result in issues being pressed or not; and level 4, actions that influence whether or not an issue is even formulated. While pluralists claimed that only the first level could be studied empirically, postplural-

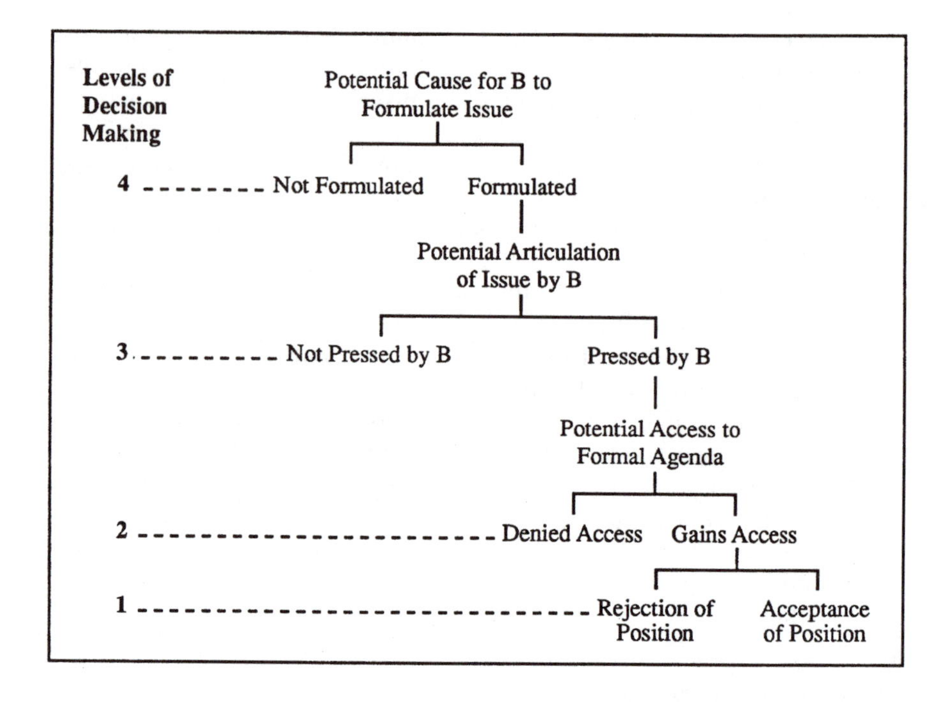

Figure 1. Levels of decision and nondecision making. Adapted from Peter Saunders, *Urban Politics: A Sociological Interpretation* (London: Hutchinson, 1979), p. 29.

ists argued that the higher levels at which various forms of "nondecision making" (that is, informal decisions affecting such questions as the selection of policy arenas, the composition of committees, and the organization of agendas) occurred could be far more influential. Decisions at these levels represented a "second face of power" ignored by pluralists.

When pluralists argued in response that what could not be measured did not exist (a classic behavioralist move), Bachrach and Baratz conceded that conflict was necessary to reveal the operation of power beyond the formal arena. That concession effectively restricted their analysis to levels 2 and 3. In contrast, Lukes claimed that the most "effective and insidious" exercise of power occurred at level 4, when values and perceptions were manipulated in such a way that people failed even to formulate needs or demands. His argument was based on a further claim, that people have interests that they may not be free to recognize or formulate but that may be revealed, especially during periods of radical social transformation.[6]

Thus the pluralist-postpluralist debate generated a further important problem: that of defining and investigating interests. While pluralists insisted on equating interests with "preferences" expressed at level 1, postpluralists proposed a realist definition of interests as values and goals revealed under conditions of autonomy.[7] But this definition is fraught with further problems: it produces a temptation to formulate a universal conception of freedom and it promises an infinite regress of essentially contested concepts. Fortunately, there is a third position. Instead of asking what interests *might* be under ideal conditions, one can ask what they *are* or *have been* in specific social and cultural circumstances. Such a constructivist position assumes that interests are contingent and that they can be defined as part of a historical inquiry into the formation of the values, beliefs, and practices of institutions and groups.

Poststructuralist treatments of power have denied the core voluntarist assumption of the conscious and responsible actor, operating in central locations. According to Foucault, the analyst should ask not, "Who has power and what does he have in mind?" but, "[How is power expressed and experienced] at its extremities, in its ultimate destinations, . . . those points where it becomes capillary, that is, in its more regional and local forms and institutions?" Foucault's achievement was to demonstrate the ways in which discursive practices incorporate and transmit power relations by supporting a "normalizing gaze"—a socially defined system of rules that permits certain statements to be made, orders these statements, and allows us to identify some statements as true, others as false, and still others as irrelevant.[8] He demonstrated the power-ladenness of language in a manner that parallels Thomas S. Kuhn's analysis of the theory-ladenness of observation.

On the face of it, the postpluralist and poststructuralist traditions seem to conflict, the former insisting on the primacy of the active, responsible subject, the latter insisting on denying the subject. At least, this is the received wisdom. But this conflict is not necessary. Indeed, Foucault in some passages seems to acknowledge this. He writes in *The History of Sexuality* that in analyzing a political situation "it is *often* the case that no-one is there to have invented [power relations], and few who can be said to have formulated them."[9] Such an observation leaves open the possibility of analyzing the creations of the inventors and formulators in other situations. The active and capillary forms of power are not contradictory types, and there is no reason why we cannot study both forms. Why not put the subject back, where many of us think it belongs, in the poststructuralist landscape?

Such a synthesis enriches both postpluralist and poststructuralist approaches. First, it supports examination of the social and cultural contexts in which governmental and sectoral interests in a contested resource like genetic engineering are formed. Second, it guides examination of how these interests operate in "mobilizing bias," not only when shaping the policy arena, as postpluralists have proposed, but also when establishing discursive practices. And finally, it directs attention to the ways in which discursive practices are both shaped by and reinforce power. Indeed, the process of establishing a specific discourse may be expected to reveal a great deal about the politics of a given policy arena. This synthesis makes it possible to pursue lines of inquiry that poststructuralism represses: to make connections between power at its extremities and at its points of initiation, and to ask how institutions and discursive practices become established. This opens up a response to an acute criticism of poststructuralism, that it re-levels the playing field by implying that anyone or anything can have power.[10] In asking who is doing the shaping of the policy arena, and how, and at what levels, this approach allows one to assess the strengths of the various interests being exercised. In summary, one can have one's poststructuralist cake and eat it postpluralistically.

Applying this approach to the history of genetic engineering policy challenges two widely held interpretations. The dominant view, *technical reductionism*, assumes that the problems posed by genetic engineering were solved through a rational process of assessment conducted by technical experts. Accordingly, the dismantling of the American and British genetic engineering controls is portrayed as the achievement of technical experts with privileged access to specialized knowledge in the face of resistance from an irrational public.[11] A second position attributes the rise and fall of the controls to *domination by an international scientific elite*. According to this view, this elite "closed ranks" on the concerned public, acting to persuade the latter that the field was safe but aware that definitive evidence for this claim did not exist.

Certainly one can find some support in the available evidence for both positions. But there is much that they cannot explain. I want to argue that the methods indicated above generate challenges to both views and support a third, more complex and dynamic picture of the operation of socioeconomic interests in genetic engineering, of power in both its initiatory and capillary forms, and provide criteria for deciding which interests were ultimately most significant. This is an interactive, multidimensional approach that allows the analyst to remain open to the actual complexity of struggles over science and to assess the strengths of the various interests being exercised.

Phase 1, 1972–1975: Problem Definition

In the first phase, policy making was an inside affair, shaped by those close to the field. Committees were established to examine the problem: in the United States, the National Academy of Sciences committee chaired by Paul Berg, and later the Recombinant DNA Advisory Committee (RAC) of the National Institutes of Health (NIH); in the United Kingdom, the Working Party on the Experimental Manipulation of the Genetic Composition of Micro-organisms convened by the Advisory Board for the Research Councils under the chairmanship of Lord Ashby. Their memberships were drawn almost exclusively from the communities of molecular biologists within which the techniques were being developed and from government and private organizations concerned with the advancement of biomedical research.

In 1974–75 other sectors had not yet engaged the issue. The private sector was certainly alert to the industrial potential of genetic engineering, but while the field was in its infancy and its long-term implications remained uncertain, corporations were content to keep watching briefs on developments in both the scientific and the political arenas. An internal memo on a meeting in 1975 of executives of the British multinational company, Imperial Chemical Industries (ICI), with representatives of a small genetic engineering firm records ICI's response: "To invest money to achieve expression of mammalian genes in *E. coli* would be worthless because the likelihood is that this will first be done by one of the many academics working in this area; then we will all have access to the information.[12]" Similarly, industry watched political developments, saying and doing little at this stage.

The larger public was not yet alert to the social and environmental issues raised by the field. Public interest organizations in the United States and trade unions in the United Kingdom would become involved later. So at this point (1973–75), the main players in both countries were the scientific communities and the government agencies responsible for supporting genetic engineering. In both countries, however, strong public concerns about environmental pollution and workplace health and safety had recently emerged, and in that social context the unrestrained deployment of science and technology might well be challenged. Scientific leaders in each country recognized that possibility. At a conference organized by the CIBA Foundation in London in 1971 and attended by British and American scientific leaders—including two future chairs of British genetic engineering policy committees, Sir Gordon Wolstenholme and Lord Ashby—Alan Bullock, vice-chancellor of Oxford University, posed the problem as follows:

> If this [critical] attitude continues, the public will say "This is so much a Pandora's Box you are giving us, and it is so evident that it will be misused, that we had just better stop the fundamental research." I can hear people saying: "We just can't tolerate the problems that are going to be created by genetic engineering, and we will shut it down as a gift too destructive to the ordinary inventions and the ordinary *mores* of human life." . . . This is the dilemma.[13]

Thus Bullock pinpointed the problem facing leaders of science on both sides of the Atlantic. If the public were not involved in decisions on the control of science, it might very well choose to withdraw its support; if it were involved, it might move to limit or even close down branches of science. For such leaders, the Pandora's box that might be opened by genetic engineering was not a box of monsters but a box of monster regulations forced on science by a public that feared science.

The pause in research taken in both countries in 1974–75 may be seen at one level as an act of laudable self-denial. Yet those responsible for genetic engineering were quite aware that they were making preemptive strikes. Lord Ashby, chair of the first British committee to assess the implications of genetic engineering, recalled in 1977 that he felt it was important to "get [his committee's] report out quickly to clear the field of sensational statements."[14] To reassure the public, it was important not only to react to concerns about genetic engineering but to be *seen* to be acting by the respective publics in the two countries. This theme of "being seen to act," with its strong subtext of fending off public intervention, would be repeated frequently by scientists and research directors as they struggled to produce a solution to the genetic engineering problem.[15]

In 1974–75, the policy arenas established in each country—the Berg committee and the NIH RAC in the United States and Lord Ashby's working party in the United Kingdom—were largely controlled by and most responsive to the relevant scientific elites. Shaped by this mobilization of bias, discourse and policy making advanced in synergy with each other. The social problems of controlling the use of a powerful and possibly hazardous technology were transformed into the technical problem of containing the organisms used in research. Furthermore, the problem of controlling application of the new field was virtually excluded from formal consideration—a far-reaching exclusion that established a major boundary for future policy making. For example, the question of military use of genetic engineering and its possible implications was consistently kept off policy agendas.[16]

Only in one key respect did the long-term impact of genetic engineering figure in the emerging policy discourse, namely, as "benefits." This configuration of genetic engineering's implications had deep roots, particularly in the strong utilitarian tendencies in biomedical research, which stemmed from the pressure on funding agencies to justify their appropriations in terms of health benefits to society. When the question of possible hazard was raised, it seemed natural for scientists to turn to future "benefits" to justify continuing their work. Thus key events that disseminated the reductionist definition of the issue—the report of the Berg committee, the press conference on this report, and the Ashby report—opposed "benefits" of the field to its "potential hazards." In the press coverage of the genetic engineering controversy that followed, hazards and benefits would be played off against one another countless times.[17]

The reduction of the genetic engineering problem to a technical question of containment, and the virtual unanimity with which it was embraced by those with access to the chosen arenas, meant that this perception of genetic engineering would quickly become dogma. Rarely in the future would policy makers deviate from that basic position. This reduction also contributed powerfully to the policies produced in these arenas, reinforcing the central role of the respective biomedical research communities in future policy making.

The international conference held at Asilomar, California, in 1975 was organized to cement the reductionist discourse that bore within it the seeds of a technical solution, and it largely succeeded in doing so. The few dissenting voices that struggled to express concerns about possible social uses of genetic engineering or the implications of dangerous research were silenced or found those concerns transformed into elements of the technical agenda.[18] The view of some participants that some experiments were so dangerous that they should never be performed, for example, was transformed by the Asilomar negotiations into the proposal that certain risks were "of such serious nature that they ought not to be done with presently available containment facilities." Rarely in the future would policy makers stray beyond the boundaries established by the basic policy paradigm asserted and confirmed at Asilomar. The precautions proposed by the Asilomar participants were not insignificant: the conference legacy was one of caution. By the standards adopted less than a decade later, it set very high containment levels for many categories of experiment, and the burden of proof rested on scientists to show that the field was safe. But the limits set by the discursive practices affirmed at Asilomar precluded consideration of the larger social and ethical issues posed by the field.

Phase 2, 1976–1978: Policy Divergence

In the second phase of policy making, what had promised to be a straightforward implementation of the Asilomar paradigm—development of controls by technical committees—was disrupted, on the American side, by the growing public controversy that surrounded the NIH process and, on the British side, by the entry of the trade unions into the issue.

In the United States public concern erupted in controversy both locally and nationally. This concern focused on the limited scope of the NIH guidelines promulgated in 1976, which applied only to NIH-supported research but not to activities in the private sector; the virtual exclusion of the wider public from policy formation; and possible weaknesses in the controls themselves. Genetic engineering became a hot congressional issue, with over a dozen bills and resolutions introduced in 1977. Legislators introduced bills to impose universal controls, appraise the social and ethical implications of the field, and, in one case (the bill introduced by Senator Edward Kennedy), transfer responsibility for policy formation from the NIH to a freestanding commission.

The strenuous efforts of the American biomedical research community to derail the genetic engineering legislation in 1977–78 were aimed at ensuring that policy would continue to be made by the NIH. Since the NIH was the principal sponsor of biomedical research and therefore a government agency to which biomedical researchers had access, NIH control over policy formation seemed to assure scientists that the resultant policies would respond to their needs. During this period of intense controversy, the discourse concerning the genetic engineering "problem" underwent a further reduction. A series of private and semiprivate scientific meetings refocused public attention on a single argument: the impossibility of transforming the main cloning host used at that time, a strain of the common intestinal bacterium *Escherichia coli*, known as *E. coli* K12, into an "epidemic pathogen" that could escape the laboratory and spread through a population.[19]

The argument originated at a private meeting of a group of biomedical researchers held at the National Institutes of Health in Bethesda, Maryland in August 1976, as the controversy surrounding the field crescendoed. Formally, this meeting, chaired by two NIH virologists, was arranged to bring together leading biomedical researchers to design experiments to assess the hazards of genetic engineering. Informally, however, this group also moved to control the growing public debate about the field. Angered by the growing public challenge to the NIH policy, they agreed to launch a public relations campaign aimed at persuading the public that the field posed no serious risks and that experimental results would support this claim.[20]

The group's organizers focused on whether *E. coli* K12 bacteria could ever be transformed into an epidemic pathogen. As one voice on the meeting's transcript put it: "In terms of PR you have to hit epidemics, because that is what people are afraid of and if we can make a *strong* argument and make it stick, then a lot of this public thing will go away. . . . It's molecular politics, not molecular biology."[21]

That the assumptions behind this argument were misleading was understood. Epidemics are hardly the only concern about genetic engineering (harm to workers in laboratories and factories can come from first exposure); everyone knew that cloning would not be limited to the weakly *E. coli* K12 for much longer; and finally, it was understood that the argument would apply only in countries with adequate public health and sewage treatment facilities.

The "epidemic pathogen" argument was elaborated at further meetings of scientists at Falmouth, Massachusetts in June 1977 and Ascot, England in January 1978. The Falmouth meeting was organized to focus exclusively on the hazards of using *E. coli* K12 as the cloning host. The proceedings of this meeting register many questions about the capacity of genetically altered forms of the bacterium to transfer "foreign" DNA to other more robust organisms or to cause disease themselves. Indeed, the meeting reached consensus on the need for further experimental work to address these questions.[22] Nevertheless, what reached the wider scientific community and the public through the circulation of the summary report written by the conference chair was an essentially soothing view that *E. coli* K12 was incapable of pathogenic transformation. As a *New York Times* headline put it: "No Sci-Fi Nightmare After All."[23]

At the Ascot meeting, which focused on the hazards of introducing viral

DNA into bacteria, concerns were again neutralized. "If we ignore [hazards to laboratory workers], we end up looking like a bunch of virologists with a completely callous and unrealistic approach to human error," one participant commented. Nevertheless, the report emerging from the meeting ignored the question of hazards to workers, focusing exclusively on epidemic scenarios. On that basis it was claimed that the cloning of viral DNA would "pose no more risk than work with infectious virus or its nucleic acid and in most, if not all, cases, clearly present less risk."[24]

As biomedical researchers promoted these conclusions in the media, at further scientific meetings, and in the halls of Congress, they also increased their scope. By a kind of discursive mutation, the claim that cloning posed no risk of an epidemic was transformed into the far wider claim that the field posed "no extraordinary hazard."[25]

This change in perception of genetic engineering hazards was not universally accepted, especially in Britain. Indeed, British scientists viewed with skepticism the conferences that produced the argument. One scientist close to the issue later described the Falmouth meeting as "choreographed." Another stated that Falmouth "was a real set-up . . . not a comprehensive scientific debate. . . . The [epidemic pathogen argument] was developed by people who wished to produce a certain conclusion." British policy addressed occupational hazards first and foremost, so the focus on epidemics was seen as irrelevant. The British molecular biologist Sydney Brenner recalled in 1980: "We were never concerned about creating an epidemic pathogen. In the first instance, our concern was the health and safety of people at work."[26]

The same meeting in Bethesda responsible for originating the "epidemic pathogen" argument also reinforced that argument through a crucial consensus concerning the goal of empirical hazard assessment. As early as 1975, the NIH had decided to conduct an experiment aimed at exploring a worst-case scenario for genetic engineering. Bacteria that bore cancer virus DNA would be used to infect test animals such as mice, and the outcome would be monitored.

At Bethesda, two distinct experimental arrangements were considered. The worst-case approach meant using bacteria able to survive and multiply in mice. At least one of those present urged this approach, proposing *Salmonella typhimurium*, a mouse pathogen, as the bacterium to be used to carry DNA from a virus known to cause cancer in mice (polyoma virus) and to be tested on mice:

> Why not do this experiment under [very high] containment, but use a known mouse pathogen; use a *Typhimurium*. . . . Here you have an organism . . . which cuts across the gut barrier. It gets inside the epithelial cells, delivers your polyoma to an epithelial cell. . . . It goes into the liver. It goes into the spleen. It's delivering your "biohazard," and you can see what happens. . . . Take the opportunity to do a good experiment.[27]

A second approach involved using the weakly *E. coli* K12 as the test bacterium. *E. coli* K12 was not a natural mouse pathogen, however, and was known to be expelled quickly by mice, leaving little chance for long-term exposure to the viral DNA it would carry. This was rather like crash-testing a sponge-rubber dummy for auto injuries, and the Bethesda scientists knew it. One person noted "the problem with K12. You would have to do too many things to make K12 pathogenic." Others agreed: "One of the problems with K12 in the mouse is that we may have a very insensitive system for looking at changes in pathogenicity." "My concern is that if you work with K12 and you put a particular factor into it, you may clone something virulent. But you don't detect it because the strain has got how many strikes against it? It doesn't colonize. You could have a biohazard sitting in that K12 and not know it and be led to believe that it is a safe system."[28]

"Molecular politics," however, again carried the day. What was needed, the majority believed, was a "slick *New York Times* kind of experiment," one that would "capture the imagination of reporters." Risk assessment experiments would be aimed not at testing worst-case experiments but at assuring the public that hazardous genes could be contained in the weakly *E. coli* K12.[29] This is a fundamental reason why the received view that the hazard problem was solved on technical grounds is invalid: the worst-case experiments that might have provided relevant evidence were not performed. Virtually all the experiments completed tested the goal agreed on at the Bethesda meeting: the ability of *E. coli* K12 to contain novel genes.[30]

While U.S. scientists successfully defended their chosen policy arena, the NIH, British scientists ceded important ground to the trade unions that represented technical staff, including scientists.[31] In Britain, a compromise emerged between the interests of the research establishment in pursuing genetic engineering under voluntary controls and those of the unions in maintaining laboratory safety. (As in the United States, private industry in Britain maintained a low profile.) A committee chaired by Sir Robert Williams was charged with developing a framework for implementing the Ashby working party's recommendations. The outcome of the Williams committee's consultations with various scientific organizations, trade unions, and industry representatives in 1975–76 was a decision to regulate genetic engineering under the Health and Safety at Work Act of 1974. The Health and Safety Commission (comparable to the U.S. Occupational Safety and Health Administration) would implement regulations; the Medical Research Council would provide the secretariat—in other words, would mediate the flow of advice to the Health and Safety Commission. And ambiguously positioned between the two institutions was an anomalous creature known as "GMAG"—the Genetic Manipulation Advisory Group—formally existing outside both agencies as a QUANGO (quasi-autonomous nongovernmental organization).

A broadly constituted committee with representation of the scientific, business, and labor sectors as well as the vaguely defined public, GMAG was sometimes referred to as an "experiment in social democracy," the brainchild of the Secretary of State for Education and Science, Shirley Williams. But it was more (or less) than that, for the government bureaucracies that created it expected it to function within the framework developed by the Ashby and Williams committees. Thus GMAG could not simply balance conflicting interests: it resolved conflict largely within a preexisting framework of assumptions and practices. Despite that qualification, GMAG provided the trade unions with a significant voice in policy making, especially on safety issues.

As a result, for about two years (1977–78), the U.S. and British policies diverged. The British government enacted a regulatory policy that applied universally. The American government retained the voluntary controls developed by the NIH. These did not cover research, development, and application of genetic engineering by the private sector, although the latter claimed that they complied voluntarily. And while the British committee operated conservatively, within the framework established by the Williams working party, its counterpart in the United States, the RAC, began to prepare for the first major revision of the controls developed in 1976.

If the British government accepted the need for government regulation of genetic engineering, what can explain the success of the American biomedical research community in resisting efforts by members of Congress to enact a similar regulatory system in the United States? Admittedly, American scientists launched a sophisticated and extraordinarily energetic campaign against the legislative proposals, especially that of Senator Kennedy. But Congress has been tough with science under other circumstances. What made genetic engineering different from other powerful research techniques was its commercial potential. By the fall of 1977, the field had provided a first concrete indication of that potential—the expression of the somatostatin gene in *E. coli*.[32] That event figured prominently in hearings before the Senate Subcommittee on Science, Technology, and Space chaired by Adlai Stevenson in November 1977, where scientific leaders hailed the experiment as signifying "extraordinary progress towards the construction of organisms that make therapeutically useful hormones."[33] At this point, the ground of the genetic engineering debate shifted rapidly. Health risks receded in importance, defused by the arguments constructed at the Bethesda and Falmouth meetings. The "real" risk was now defined as that of losing out on a novel field with immense commercial impact. As Philip Handler, president of the National Academy of Sciences, stated at the Stevenson hearings: "Cumbersome and punitive legislation is not needed. The financial cost of overly cautious containment and enforcement, the delay in achieving benefits, and the penalties incurred by restricting freedom of inquiry are real risks to be considered in setting up regulations." And Joseph Stetler, president of the Pharmaceutical Manufacturers Association (PMA), brought the full weight of the transnational members of the PMA behind that position: "It is quite possible that legislation could be so restrictive, so much of a disincentive, that our people wouldn't lose their interest, their curiosity . . . they would go overseas."[34]

By the spring of 1978, this argument figured as one of the principal conclusions of the report of the House Subcommittee on Science, Research, and Technology. As Daniel Kevles notes in this volume, by this point the "commercial prospects of biotechnology figured prominently in the minds of federal policymakers." Citing the testimony of John Adams, the vice president of the Pharmaceutical Manufacturers Association, the report warned: "Any significant U.S. lag in genetic technology, for whatever reasons it may occur, could result in diminished power and prestige for this nation on the international scene."[35] Thus the structure of international competition in science entered as a major factor in shifting congressional interest from regulation of genetic engineering to its promotion. If genetic engineering had provided merely a set of powerful and risky research techniques without also offering a basis for new forms of production, it seems unlikely that members of Congress could have been persuaded by slight scientific arguments or even energetic lobbying to drop their interest in regulation.

It was at this stage, then, that the influence of the private sector, especially that of major multinational corporations, began to be felt. While industry representatives were not nearly as visible as the scientific sector in the congressional debate (they hardly figure in news stories from this period), it would be a mistake to equate their invisibility with inaction or with lack of influence. Industry representatives were consulted by government officials on both sides of the Atlantic. In June 1976, before issuing the first NIH guidelines, the NIH director met with industry representatives—and was informed, bluntly, that they preferred voluntary compliance, and if possible, no regulation by the federal government. The industry representatives then noted a number of problems they had with the NIH guidelines, especially their "overriding concern" over the protection of commercially sensitive industry information—an issue that would figure prominently in the next phase of policy making. Industry representatives were also consulted by the mastermind behind the scientists' campaign against legislation, Harlyn Halvorson, a professor of biology at Brandeis University and president of the American Society for Microbiology. Halvorson's phone log reveals a network of alliances that encompassed industry, science, and NIH officials, and it shows that industry officials, while not wanting "to tangle" with the DNA issue, quietly supported Halvorson's legislative strategy.[36]

But the most influential actions taken by the private sector in this period were not related directly to struggles over and within the policy arena. Rather, they shaped interests, practices, and the social climate in which future policy would be formed. First, from the demonstration of the bacterial expression of the somatostatin gene in 1977 onward, multinational cor-

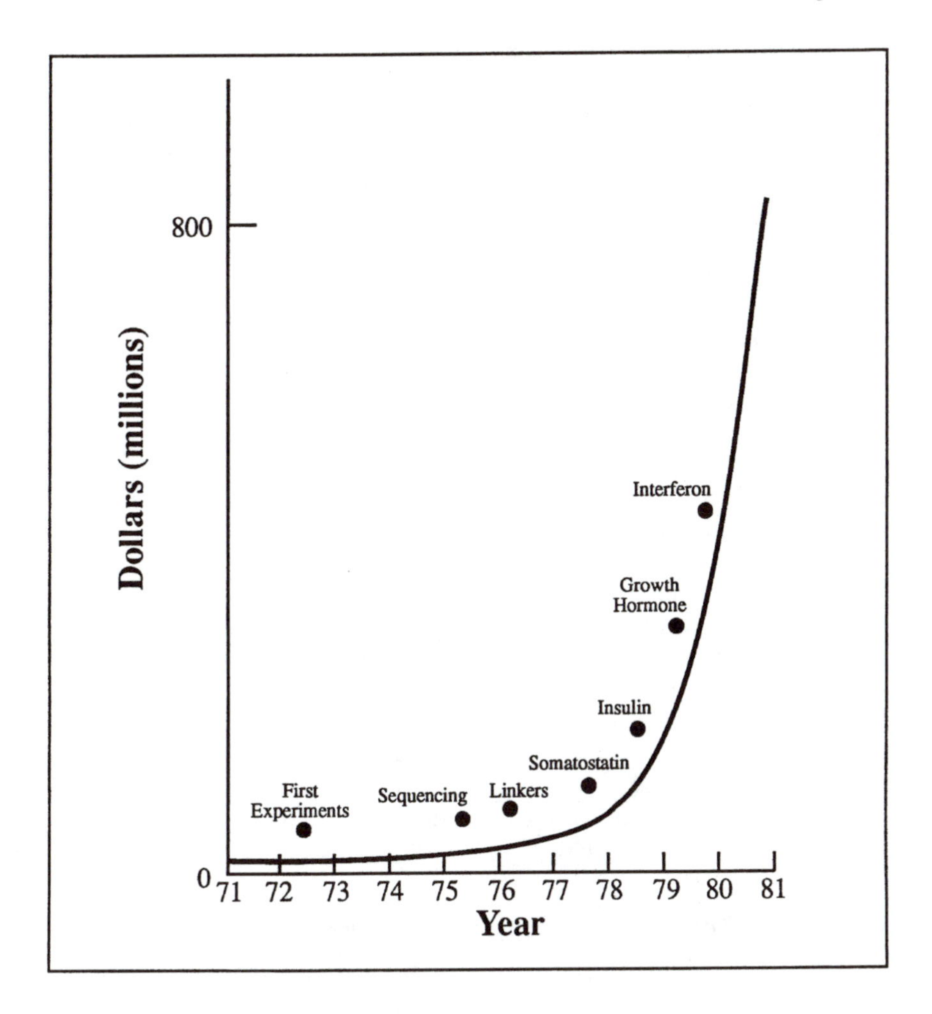

Figure 2. Equity investment in biotechnology companies, 1971–81. Adapted from Leslie Glick, "The Biotechnology Industry," paper for the American Law Institute and the American Bar Association, Committee on Continuing Professional Education, October 1981.

porations and venture capital firms began to invest heavily in small genetic engineering firms (Figure 2). Corporations also formed joint contracts with genetic engineering firms and universities, established large in-house research programs, and acquired small firms. These investments initiated the field of molecular biology into a new commercial role. Members of the field, formerly cloistered in academe, became equity owners, corporate executives, and industry advisers. Of course, ties between academe and industry were not new. What *was* new was the extent of these ties and the intensity of their effects.[37]

Second, throughout the 1970s the American business sector worked to transform the social relations of science and technology—first, through achieving social and economic conditions favorable to rapid technological development; second, through shaping a political climate favorable to deregulation.[38] Toward the end of Carter's term, the business community was well on the way to achieving both goals. By now the government had begun to subject regulatory policy to economic analysis and had passed measures to support industrial innovation and to foster university-industry cooperation. One such measure was the 1980 Patent and Trademark Amendment Act, which gave universities, small businesses, and nonprofit organizations rights to patents arising from federally supported research—a major government action that facilitated industry access to university research. The Reagan administration, like its predecessor, was strongly committed to facilitating rapid technological development, but unlike its predecessor, it had no obligations to labor unions, environmentalists, and consumer advocates. Consequently, it combined further incentives for technological development with a zealous deregulatory policy. Perhaps the most significant was the 1981 Economic Recovery Tax Act, which simultaneously provided substantial tax credits for research and development and reduced the long-term capital gains tax—moves that encouraged a flood of investment in biotechnology in the early 1980s. In all these ways, the business sector was enormously influential in transforming the socioeconomic context in which genetic engineering policy would be made from 1979 onward—and in shaping social interests in the field.[39]

Phase 3, 1979–1982: Policy Convergence

By the beginning of the third phase considered here, the political climate was changing rapidly from support for control of risky fields of science and technology to support for deregulation. (Congressional behavior in response to the genetic engineering issue may be seen as a bellwether.) At the same time, a new alignment between academic science and industry was being forged by government policies.

The fate of the American and British genetic engineering controls at this point is well known in outline: a precipitous rush to weaken controls in the United States, which was followed in the United Kingdom (Figure 3). However, the questions of how these policy changes were achieved and which sectors influenced them have not been addressed. Both prevailing interpretations—technical reductionism and the influence of an international scientific elite—fail to account for important features of these developments.

Consider the main interests in genetic engineering by 1979. First, the pharmaceutical and chemical industries were investing heavily in the field,

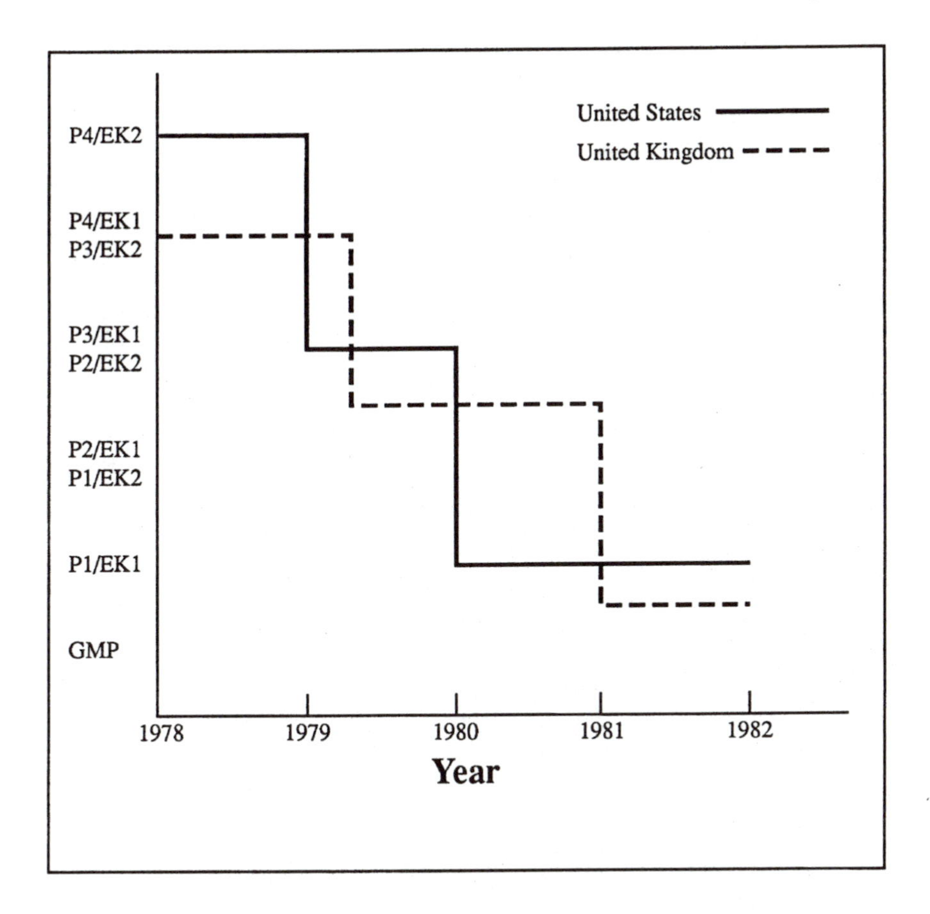

Figure 3. Changes in U.S. and British requirements for cloning the human insulin gene in *E.coli* K12, 1978–82. P1—physical containment requiring standard techniques used in microbiological laboratories. P2—P1 containment with additional separation from surrounding facilities. P3—high physical containment, using negative air pressure and laminar flow hoods. P4—extremely high containment in a specially constructed facility; used to handle the most lethal pathogens. EK1—biological containment using *E.coli* K12 as host organism. EK2—biological containment using a specially weakened strain of *E.coli* K12, able to survive only under certain conditions. GMP—good microbial practice, a British standard slightly less rigorous than P1.

and intense races for patents and products were under way. The NIH controls were troublesome for the private sector for several reasons. At the beginning of 1979, containment requirements were still relatively high, and high containment increased production costs. This was a particularly serious concern for the new genetic engineering firms like Genentech that could ill afford delays and high overhead costs.[40] Multinational corporations were probably less concerned about the costs of containment. But, as noted earlier, they strongly resisted the requirement that they divulge commercially sensitive information to the RAC, an issue they raised with the NIH director as early as 1976. At five further meetings with officials from the Department of Commerce, the Department of Health, Education, and Welfare, and the NIH, held in 1976–79, they repeated this concern, setting forth their vision of a compliance scheme: registration of research and disclosure of sensitive commercial information (that is, details of modified organisms and genes) only at the local level, to committees appointed by the industries themselves.[41] In the short term, however, they had to settle for some disclosure to the RAC, and negotiations focused on how much and under what conditions.

The second strong interest in NIH policy was that of academic scientists. Bound by the NIH guidelines, they now felt seriously disadvantaged by the double standard that characterized implementation of the NIH controls. Whereas the NIH had no legal authority to compel companies to comply with the guidelines, it could withdraw support from deviant NIH grantees. While their counterparts in commercial laboratories could move quickly with experiments, university researchers were often delayed by NIH requirements for RAC review and approval. What university scientists wanted, therefore, were lower containment and oversight requirements so that delays would be avoided and proposals screened only at the local level.

The third interest in the NIH controls was that of environmental organizations and trade unions. These sectors wanted a policy that was cautious, participatory, and applied universally, but the congressional debacle made such aspirations appear quixotic. The government's only concession to these interests was to expand the RAC to include a few public interest representatives.[42]

How were these interests expressed and how did the NIH and its advisory committee respond? It should be remembered that despite the addition of the public interest members, the RAC continued to be defined as a technical committee, and a large majority of its members were research scientists. There was no direct industrial representation, and in 1979 only three or four members had formal ties with industry (although it is likely that other members anticipated forming such ties).[43] Given this composition, it might be expected that NIH policy would have favored academic research interests, but this was not always the case.

Where academic and industry interests converged, their effects on and within the RAC were dramatic. Until 1979, the NIH and its advisory committee had worked within the policy framework elaborated at the Asilomar conference, implementing relatively strict containment and oversight requirements. But once it was clear that congressional interest in regulation could be blocked, the scientific community, with the quiet support of industry, was well positioned to achieve a policy reversal. From 1979 onward, the discourse and practice of "containment of unknown hazards" gave way to the claim that genetic engineering posed "no extraordinary hazard," and practices soon matched rhetoric. The drops in containment requirements were so precipitous that they might almost be described as free-fall. In the absence of an effective countervailing force operating inside or outside the RAC, the practices established at the Asilomar conference (containment of unknown hazards) were swiftly abandoned as the burden of proof shifted from those who wished to pursue genetic engineering to those who wished to control it. By 1982 controls had been all but dismantled and the entire spectrum of living things opened up to genetic manipulation.

Given the convergent interests of science and industry in achieving this policy reversal and the former's strong representation on the RAC, these decisions are understandable. More interesting, because it provides a litmus test of the politics of the NIH, is the agency's response when the interests of academic science and industry diverged, most notably on the question of the conditions under which industry would voluntarily comply with the NIH controls. The specific proposal before the RAC in 1979 (the outcome of private NIH consultations with industry) was that the committee would review industry proposals in closed session, with broad protection of corporate information, including criminal penalties for unauthorized release of such information.

On this issue the committee was seriously divided, providing its public interest members with a rare opportunity to exert some influence on the outcome. A widely held view on this predominantly academic committee was that the NIH should not be transformed into a quasi-regulatory agency. Rather than accept the proposal, the committee advised the NIH director Donald Fredrickson, by a close vote, that *mandatory* controls were needed—implying that these should be developed by some other branch of government. In response Fredrickson chose to play an active role: skillfully working to overcome the committee's "malaise" without triggering congressional intervention; narrowing the committee's choices to alternatives acceptable to industry; warning the committee that nothing less than the "maintenance of parity among the nations for safe use of [genetic engineering] and access to its benefits" was at stake, and eventually extracting the advice industry desired—weak controls with strong protection for trade secrets.[44] A telling characteristic of the NIH policy shaped in the early 1980s was that the only *criminal* penalty was for release of proprietary information, not the release of the genetically engineered organisms whose dissemination the controls were supposed to restrain. Where academic and industry interests conflicted, RAC managers proved more responsive to the needs of the pharmaceutical industry than to those of the academic scientists who were the primary NIH clients.

The British response to the precipitous dismantling of the NIH controls further reinforces the argument that the reversal of the Asilomar paradigm was accomplished for political and economic, not technical, reasons. By the beginning of 1979 the British controls, influenced by union commitments to laboratory safety, were significantly stricter than those in the United States (Figure 3). A member of GMAG recalled the atmosphere: "There was an element of real panic here. . . . People will put up with being marginally out of line but not [to that extent]." Furthermore, at a meeting held at Wye College in England and jointly sponsored by the British Royal Society and the Committee on Genetic Experimentation (COGENE), a private non-governmental scientific organization, NIH officials and prominent American biologists left no doubt about their intention to weaken the NIH controls even further. The NIH director, for example, predicted the "eventual elimination" of the NIH controls.[45] The British government was now pressured by leaders of industry and science to follow the American lead. The press carried reports of scientists leaving for countries with looser controls. Industry officials warned that Britain was about to lose a "vital new heartland technology." Further pressure was exerted by the influential European Molecular Biology Organization, which in December 1978 endorsed the American controls.[46]

Under tough economic conditions in which British preeminence in science was perceived as a major economic resource, governmental responsiveness to the calls from industry and science transcended party affiliations: it was James Callaghan's Labour government, not the Conservative government of Margaret Thatcher, that took the first actions to weaken Britain's regulatory policy. The Callaghan government established a high-level committee chaired by Alfred Spinks, a former ICI executive, which soon produced a "hard-hitting, interventionist" report recommending vigorous promotion of biotechnology and weakened controls for genetic engineering.[47] In addition, the Labour government appointed Sir William Henderson, whose commitment to facilitating rapid development of genetic engineering was well known, as the new chair of GMAG. Thatcher's Conservative government went further, undercutting the principal force favoring caution, the trade unions, reducing support for the Health and Safety Commission, and conveniently swallowing free market scruples to endorse governmental promotion of the field. A government white paper on biotechnology that embodied a free market philosophy, arguing that industry, not the government, should be responsible for transforming genetic engineering into a commercially viable technology, was quietly shelved.[48]

A stream of further reports issued from the House of Lords, the House of Commons, the Royal Society, and elsewhere, calling for government support for commercial biotechnology, selective funding of research, educational initiatives aimed at producing a technical infrastructure, and less restrictive controls.[49] In December 1980 a government-sponsored genetic engineering firm, Celltech, was created with capital from the British Technology Group, the Technology Development Corporation, and private sources and involving a collaborative arrangement with the Medical Research Council. By 1981 it was estimated that the British government was spending on the order of £25–30 million a year on biotechnology.[50]

Under these conditions, GMAG's commitment to maintaining caution waned. As one GMAG member expressed the change:

> I think it is hard to convey . . . the extent to which there was a shift, a gearing down, in the level of activity invested in [controls], from 1978, when everything went to GMAG, when there was an enormous amount of interest and concern in this area, when local safety committees met regularly, and when the entire system functioned at a very high profile, to the situation in 1982 when there was *no* interest in pursuing health and safety aspects of this area. . . . Anyone trying to maintain a high profile role for GMAG . . . was unable to do so. Basically, GMAG was at every stage of the game progressively being ignored, and since it did not have political clout, there was no way to counter that. The balance of forces [affecting operation of the committee] shifted totally.[51]

Driven at each stage by concerns about being "left out on a limb" in comparison with controls elsewhere, GMAG struggled to regain parity. A risk assessment scheme developed for GMAG in 1978 by Sydney Brenner and others proved useful in this regard. Each time the NIH containment requirements plunged lower, key variables of the scheme were adjusted, bringing the British requirements into line (Figure 3).[52] With respect to large-scale processes, industry representatives on both sides of the Atlantic played not only a general role in pressing for deregulation but also a specific one, lubricating the downward slide of the controls for industrial-scale processes with predictions of imminent relaxation on the other side of the Atlantic.[53] The fear of being placed at a competitive disadvantage by weaker regulations elsewhere proved a powerful catalyst for deregulation. As in the second phase of policy formation, the structure of international industrial competition provided substantial leverage for the arguments of scientists, industrialists, and government bureaucrats.

Concluding Appraisal

In summary, there is no simple, one-dimensional explanation of the rise and fall of genetic engineering controls. The dominant view that policy makers solved the problem rationally fails both on the broad grounds that the problems posed by genetic engineering were not merely technical in nature and on the narrower grounds that the evidence for assessing hazards was not only unavailable but *designed* to be unavailable for political reasons. The second view described above — domination by an international technical elite — certainly illuminates a necessary condition for policy formation. Technical elites were highly influential in shaping the initial policy discourse (reduction of the problem to a technical one) and in generating later arguments that dismissed the biohazard problem altogether. But this view cannot explain the reshaping of the socioeconomic climate in the early 1980s nor the reshaping of the interests of the technical elite itself. Nor, as I have indicated above, can it explain important features of policy formation, especially those of the second and third phases. Had genetic engineering been merely a powerful research tool—had it not also provided the basis for a new form of industrial production—it seems likely that the scientific community's rationales for deregulation would have met with far greater skepticism. Examination of the socioeconomic contexts of policy making, the formation of social interests in genetic engineering, and the operation of these interests reveals the pervasive influence of the private industry—in shaping a political climate favorable to deregulation, in forging an alignment between academic science and industry, and, when necessary, in pressuring government agencies on particular aspects of policy. No other sector operated with the same range or clout.

But such an emphasis on human agency does not preclude acknowledging the importance of the competitive structures of the integrated global economy in which contemporary science and industry are embedded. Individual scientists, scientific organizations, small genetic engineering firms, and large multinational corporations could exert leverage in the policy arena by virtue of those competitive structures. The greater the amount of financial and intellectual capital that an organization could move across national boundaries, the greater that organization's influence in a national arena.

These structures also limited the capacity of individuals and organizations to act differently. The structure of global industrial competition, transmitted by corporations that are free to move personnel and capital across national boundaries, led states to compete to attract and keep for new sources of innovation, industry, and employment. Leaders of science and industry and heads of government agencies pursued their interests with respect to genetic engineering controls, but they did so under socioeconomic conditions that were not theirs to choose.

Of course, these competitive structures are not necessarily permanent fixtures on the global landscape, nor are interests and policies completely determined by them. Nations can and do act to contain competition in the industrial and military spheres through environmental, trade, and arms control agreements. The present limits to the effectiveness of these agreements are obvious, but should not blind us to their surprising successes in some areas. Risky technologies like genetic engineering pose a long-term political challenge: how to devise new forms of international cooperation that develop these technologies to respond not to competitive imperatives but to basic human need.

The research on which this article is based was supported by a Rockefeller Foundation Humanities Fellowship and by National Science Foundation grants SES7826618 and SES8511131.

Notes

1. This essay draws on more detailed arguments and empirical evidence given in several previous articles and in Susan Wright, *Molecular Politics: Developing American and British Regulatory Policy for Genetic Engineering, 1972–1982* (Chicago: University of Chicago Press, 1994).

2. For discussion of this tradition, see Steven Lukes, "Power and Structure," in Lukes, *Essays in Social Theory* (New York: Columbia University Press, 1977), pp. 1–29.

3. Hubert Dreyfus and Paul Rabinow, *Michel Foucault: Beyond Structuralism and Hermeneutics*, 2nd ed. (Chicago: University of Chicago Press, 1983), pp. 143–204.

4. See, e.g., Robert Dahl, *Who Governs? Democracy and Power in an American City* (New Haven, Conn.: Yale University Press, 1961).

5. See especially Peter Bachrach and Morton Baratz, "The Two Faces of Power," *American Political Science Review* 56 (1962), 947–52; and Bachrach and Baratz, "Decisions and Nondecisions: An Analytical Framework," *American Political Science Review* 57 (1963), 641–51.

6. Peter Bachrach and Morton Baratz, *Power and Poverty: Theory and Practice* (New York: Oxford University Press, 1970), pp. 49–50; and Steven Lukes, *Power: A Radical View* (London: Macmillan, 1974), pp. 23–24.

7. William Connally, "On 'Interests' in Politics," *Politics and Society* 12 (1972), 459–77; and Lukes, *Power*, pp. 34–35.

8. Michel Foucault, *Power/Knowledge: Selected Interviews and Other Writings, 1972–1977* (New York: Pantheon, 1980), pp. 96, 211.

9. Michel Foucault, *The History of Sexuality*, vol. 1, *An Introduction*, trans. Robert Hurley (New York: Pantheon, 1978).

10. For trenchant criticism of this problem, see Nancy Hartsock, "Foucault on Power: A Theory for Women?" in *Feminism/Postmodernism*, ed. Linda J. Nicholson (New York: Routledge, 1990), pp. 157–75, on pp. 168–70; and Steven Shapin, "Following Scientists Around: Review of Latour, *Science in Action*," *Social Studies of Science* 18 (1988), 533–50.

11. The *locus classicus* for this position is an unpublished paper by former NIH director Donald Fredrickson, entitled "Science and the Cultural Warp: RDNA as a Case Study," given in 1982 at the annual meeting of the American Association for the Advancement of Science in Washington, D.C.

12. Imperial Chemical Industries, memo, 2 July 1976, Imperial Chemical Industries press files, Runcorn, United Kingdom.

13. Quoted in Gordon Wolstenholme and Maeve O'Connor, eds., *Civilization and Science: In Conflict or Collaboration?* (Amsterdam: Elsevier, 1972), p. 121.

14. Interview with Eric Lord Ashby, by Susan Wright, 10 Jan. 1977.

15. See, e.g., Nicholas Wade, "Genetics Conference Sets Strict Controls to Replace Moratorium," *Science* 187 (1975), 931–35, on p. 935, citing British molecular biologist Sydney Brenner.

16. For details see Susan Wright, "Biotechnology and the Military," in *Agricultural Bioethics: Implications of Agricultural Biotechnology*, ed. Steven M. Gendel et al. (Ames: Iowa State University Press, 1990), pp. 79–96.

17. Paul Berg et al., "Potential Biohazards of Recombinant DNA Molecules," *Proceedings of the National Academy of Sciences* 71 (1974), 2593–94; and *Report of the Working Party on the Experimental Manipulation of the Genetic Composition of Microorganisms*, Cmnd. 5880 (1975). See also Robert Bud's article in this volume.

18. Paul Berg et al., "Asilomar Conference on Recombinant DNA Molecules," *Science* 188 (1975), 991–94, on p. 994.

19. Susan Wright, "Molecular Biology or Molecular Politics? The Production of

Scientific Consensus on the Hazards of Recombinant DNA Technology," *Social Studies of Science* 16 (1986), 593–620; and Wright, *Molecular Politics* (note 1), pp. 221–55.

20. National Institutes of Health, Office of Recombinant DNA Activities, Enteric Bacteria Meeting, 31 August 1976, transcript. For excerpts, see Wright, *Molecular Politics* (note 1), pp. 457–64.

21. NIH, Enteric Bacteria Meeting transcript (note 20), p. 45; and Wright, *Molecular Politics* (note 1), p. 461.

22. Sherwood L. Gorbach, ed., "Risk Assessment of Recombinant DNA Experimentation with *E. coli* K12," *Journal of Infectious Diseases* 137 (1978), 613–714.

23. Sherwood Gorbach to Donald Fredrickson, 14 July 1977, NIH Office of Recombinant DNA Activities records; "No Sci-Fi Nightmare After All," *New York Times*, 24 July 1977.

24. National Institutes of Health, U.S.-EMBO Workshop to Assess the Containment Requirements for Recombinant DNA Experiments Involving the Genomes of Animal, Plant, and Insect Viruses, 27–29 Jan. 1978, transcript, p. 1061; and National Institutes of Health, "U.S.-EMBO Workshop to Assess Risks for Recombinant DNA Experiments Involving the Genomes of Animal, Plant, and Insect Viruses," *Federal Register* 43 (31 March 1978), 13748–55.

25. For details see Wright, *Molecular Politics* (note 1), pp. 251–55.

26. Interview with Derek Ellwood, 21 April 1980; interview with a member of the Genetic Manipulation Advisory Group, April 1980; interview with Sydney Brenner, 6 May 1980; all by Susan Wright.

27. NIH, Enteric Bacteria Meeting transcript (note 20), pp. 52–53; Wright, *Molecular Politics* (note 1), 462.

28. NIH, Enteric Bacteria Meeting transcript (note 20), pp. 59, 63, 65; and Wright, *Molecular Politics* (note 1), pp. 463–64.

29. NIH, Enteric Bacteria Meeting transcript (note 20), pp. 52, 59; and Wright, *Molecular Politics* (note 1), pp. 461, 463.

30. As it turned out, the results of several experiments were themselves controversial. For details see Sheldon Krimsky, *Genetic Alchemy: The Social History of the Recombinant DNA Controversy* (Cambridge, Mass.: MIT Press, 1982), pp. 244–63; and Wright, *Molecular Politics* (note 1), pp. 246–51, 364–80.

31. Wright, *Molecular Politics* (note 1), pp. 195–212.

32. Keiichi Itakura et al., "Expression in *Escherichia coli* of a Chemically Synthesized Gene for the Hormone Somatostatin," *Science* 198 (1977), 1056–63.

33. Senate Committee on Commerce, Science, and Transportation, Subcommittee on Science, Technology, and Space, *Hearings: Regulation of Recombinant DNA Research*, 95th Cong., 1st sess., 2, 8, 10 Nov. 1977, p. 36.

34. Senate Committee, *Hearings* (note 33), pp. 16, 366.

35. House Committee on Science and Technology, Subcommittee on Science, Research, and Technology, *Report: Science Policy Implications of DNA Recombinant Molecule Research*, 95th Cong., 2nd sess., March 1978, p. ix.

36. Associate Director for Program Planning and Evaluation to Director, NIH, "Summary of Your Meeting with Private Industry, June 2," 4 June 1976, in National Institutes of Health, *Recombinant DNA Research* (Bethesda, Md.: NIH Office of Recombinant DNA Activities) 1 (1976), 423; and Harlyn Halvorson, phone log, Harlyn Halvorson personal papers, Brandeis University. For further discussion see Wright, *Molecular Politics* (note 1), pp. 266–68.

37. For details of corporate and venture capital responses to developments in genetic engineering and discussion of their effects, see Susan Wright, "Recombinant DNA Technology and Its Social Transformation," *Osiris* 2 (1986), 303–60; and Wright, *Molecular Politics* (note 1), chap. 2. For discussion of corporate and venture capital strategies, the character of the industry-university alliances, and the new roles

assumed by scientists as corporate managers and employees, see Martin Kenney, *Biotechnology: The University-Industrial Complex* (New Haven, Conn.: Yale University Press, 1986); and Kenney's article in this volume.

38. Details of the efforts of the business sector are given in David Dickson and David Noble, "By Force of Reason: The Politics of Science and Technology Policy," in *The Hidden Election: Politics and Economics in the 1980s Presidential Campaign*, ed. Thomas Ferguson and Joel Rogers (New York: Pantheon, 1981), pp. 260–312.

39. Wright, *Molecular Politics* (note 1), pp. 54–58.

40. For analysis of the pressure generated by Genentech on the NIH to reduce containment levels, see Wright, *Molecular Politics* (note 1), pp. 341–44.

41. See the following reports of various meetings of pharmaceutical and chemical industry executives with Department of Commerce officials: Victor Cohn, "Drug Industry Seeks to Alter U.S. Rules on Genetic Studies," *Washington Post*, 26 Nov. 1976, p. A3 (meeting of 19 Nov. 1976); Associate Director for Program Planning and Evaluation to Director, NIH, "Your Meeting with Representatives from the Pharmaceutical Manufacturers Association," 1 Dec. 1976 (meeting of 29 Nov. 1976); George S. Gordon (Department of Commerce), Memorandum for the Record, "Meeting to Discuss Possible Approaches to Development of Private-Sector Voluntary Compliance with NIH Guidelines for Recombinant DNA Research, Under Surveillance by the Department of Commerce," ca. Dec. 1977, Department of Commerce records (meeting of 19 Dec. 1977); excerpts from minutes, meeting of DHEW committee with representatives of the Pharmaceutical Manufacturers Association, 13 Oct. 1978, Office of the Secretary, Department of Health, Education, and Welfare records; and minutes, meeting of DHEW General Counsel Peter Libassi, DHEW, NIH, and FDA officials, and representatives of the pharmaceutical industry, 25 May 1979, Office of the Secretary, Department of Health, Education, and Welfare records.

42. The reconstitution of the RAC followed strenuous negotiations between the office of the Secretary of Health, Education, and Welfare and the NIH leadership: see Wright, *Molecular Politics* (note 1), pp. 352–54.

43. I thank Sheldon Krimsky for providing data on industry links of RAC members in 1979.

44. Recombinant DNA Advisory Committee, National Institutes of Health, minutes, 21–23 May 1979, pp. 9–10 (committee recommendation); and minutes, 5–6 June 1980, pp. 4–6 (NIH director's remarks); records of the NIH Office of Recombinant DNA Activities. For details see Wright, *Molecular Politics* (note 1), pp. 391–403.

45. Interview with Robert Williamson, 13 May 1980, by Susan Wright (quotation); and Joan Morgan and William J. Whelan, eds., *Recombinant DNA and Genetic Experimentation* (Oxford: Pergamon Press, 1979), pp. 157–60.

46. See, e.g., Confederation of British Industry, "Policy Issues Arising from Genetic Engineering," in House of Commons, Select Committee on Science and Technology, *Recombinant DNA Research: Interim Report* (London: HMSO, 1979), pp. 217–26, on p. 221; ibid., statement of Dr. B. W. Bainbridge, p. 142; and European Molecular Biology Organization (EMBO), report of the fifth meeting of the Standing Advisory Committee on Recombinant DNA, London, 2–3 Dec. 1978, EMBO files, Heidelberg, Germany.

47. Advisory Council for Applied Research and Development and Advisory Board for the Research Councils and the Royal Society, *Biotechnology: Report of a Joint Working Party* (London: HMSO, March 1980).

48. United Kingdom, *Biotechnology*, Cmnd. 8177 (March 1981); and "Scientists Slam White Paper on Biotechnology," *New Scientist*, 12 March 1981, p. 660.

49. See, e.g., House of Lords, Select Committee on the European Communities, "Genetic Manipulation (DNA)," *Sessional Papers*, sess. 1979–80, 39th Report, 4 March 1980; House of Commons, Education, Science, and Arts Committee, *Bio-*

technology: Interim Report on the Protection of the Research Base in Biotechnology, sess. 1981–82, 27 July 1982; and Royal Society, "Biotechnology and Education: Report of a Working Group," 1981, Royal Society, London.

50. "Britain's Biotechnology Company Takes Off," *New Scientist*, 13 Nov. 1980, p. 413; and "Science After the Cuts," *New Scientist*, 13 Aug. 1981, pp. 396–97. For further evidence that the Thatcher government quietly adopted the interventionist philosophy of the Spinks committee, see the article by Herbert Gottweis in this volume.

51. Interview with Robert Williamson, 7 June 1984, by Susan Wright.

52. For details of the Brenner risk assessment scheme, see Wright, *Molecular Politics* (note 1), pp. 327–34, 417–24.

53. See, e.g., Confederation of British Industry, "Issues Arising from Scale-up of Recombinant DNA," 26 March 1981; and Irving Johnson (vice president, Eli Lilly) to William Gartland, 20 July 1981; both in NIH Office of Recombinant DNA Activities Records, doc. 1027. For details see Wright, *Molecular Politics* (note 1), pp. 426–32.

The Political Economy of British Biotechnology

Herbert Gottweis

During the late 1970s European policy makers realized with great concern that their national innovation systems were not duplicating what was perceived as the successful U.S. model of industrializing biotechnology—one driven by venture capital and based on small and medium-sized companies. A discussion began about the need "to meet the challenge of the new biotechnology." By the early 1980s most European states had devised national biotechnology strategies to prepare for what was seen as a new stage in the international "high-tech race." In this essay I focus on the British state's attempts to help shape a national biotechnology industry. A study of British biotechnology policy will allow us to identify some of the potential problems of state intervention in the development of a biotechnology industry.

European countries began to construct strategies for biotechnology after the regulatory debates of the late 1970s that followed the Asilomar Conference (1975), which had focused on the risks involved in the practices of genetic engineering. Construing the new biotechnology as a "technology of the future" played a significant role in the way the controversy was framed. Furthermore, in the course of the debates, the definition given by molecular biology of the nature of hazards involved in genetic engineering prevailed over other, alternative readings. Soon the argument gained ground that the risks involved in genetic engineering were minor compared to the future benefits expected. At least for a while, until the mid-1980s, the risk debate lost momentum.

This "settling" of the debate on the ecological and socioeconomic implications of biotechnology and the establishment of a corresponding regulatory framework were important elements in the creation of the "new economic space" of biotechnology; they thus should be viewed as part of a broadly understood notion of biotechnology policy. In the past much of the literature and discussion on biotechnology policy has focused on the regulatory debates.[1] Less attention has been paid to biotechnology policy in the narrower sense of the word, namely, the state's efforts to foster the development of national biotechnology research and development by means of science and technology support policies. Efforts to foster research were especially pronounced in several European countries that lacked strongly developed venture capital industry. There, within a strikingly different system of technological innovation, state involvement took a more pronounced and more extensive form than in the United States. Rather than leaving it to the dynamics of capitalism and merely supplementing publicly funded basic research and regulatory policies—the U.S. approach as Kenney describes it in this volume—European states initiated systematic efforts. However, as I will show in the case of Britain, state intervention was plagued by a variety of difficulties and contradictions. While the initial goals were to restructure the science-industry relationship, to stimulate a varied biotechnology industry following the U.S. model, and to build a cornerstone in the newly emerging high-tech industry, biotechnology policy succeeded only in creating an economic space for established larger and multinational companies to extend their operations into the realm of biotechnology.

How can we understand this partial failure of British biotechnology policy? What were the expectations, the constitutive rationalities and definitions of the British state's intervention in recombinant DNA-related research and development? And how did representations of the British innovation system and practices of scientific research and industrial production translate into the shaping of the economic space of biotechnology as dominated by mainly the larger companies?

Any instance of policy making, such as biotechnology policy, is a complex and "messy" process involving a variety of interrelated actors, institutions, theories, ideologies, contexts, and artifacts. Policy making is in fact an endeavor to manage messiness. Textual strategies are an important organizational effort to reduce messiness, to produce political rationality, coherence, and system. I will argue that the political construction of biotechnology in Europe involved representing various elements of socioeconomic crisis, disruption, and destabilization as menaces to the collective European identity. Inscribing biotechnology as a response to a political-cultural crisis allowed the state to take an important role in defining and fixing the meaning of biotechnology in society and created a field of legitimation for technological intervention.

The politics of biotechnology in Britain is not—measured against its own goals—a success story. The gap between the policy's defined goals and its outcomes directs our attention to one of the key problems found in policy analysis: the organizational difficulties that arise when the state attempts to direct and shape institutions of civil society such as those of the economy and science.

As I show in the next section, the politics of representing biotechnology as a technology of the future and as the answer to the political-economic crisis of the 1970s created a field of legitimacy for technological intervention. Furthermore, the government tried to enframe the meaning of a British politics of biotechnology by creating a regime of signification to define the interests and to coordinate and stabilize the practices of a variety of differential actor positions configured in the institutional complexes of the innovation system, such as the research and development (R&D) system and industry. But, as I discuss in the section after, the government's commitment to promoting biotechnology was subverted by the very construction of the neoliberal strategy adopted to promote it. Because the institutional logic of the British system of innovation implied that no political space existed for negotiating conflicts between the agencies involved in technology policy, the chosen politics of biotechnology failed to have the envisioned impact.

Inscribing Biotechnology as a Future Technology

When European governments began to devise biotechnology strategies in the early 1980s, these strategies were governed by a core assumption: that dramatic breakthroughs had recently occurred in molecular biology. The shaping of molecular biology as a new discursive formation interconnecting scientists, theoretical concepts, technological practices, and institutions at the interface of biology, physics, and chemistry dates back to the 1930s. Not until the 1960s, however, did this gradually emerging new system of signification for appropriating and redefining a wide range of disciplines from virology to biochemistry begin to be inscribed in rationalities of governmental decision making in Europe.[2]

British molecular biology became an object for governmental decision making in the early 1960s. With the development of recombinant DNA technologies this process of overseeing scientific practices, delineating new priorities of state intervention, accelerated further. Since the early 1970s, in the field of recombinant DNA in particular, new technologies were completely changing what used to be the "old" biotechnology. In Britain as well as in other countries the "new" biotechnology was targeted by policy makers and scientists for promotion as a field of crucial scientific and socioeconomic importance.

Within the framework of the relatively centralized systems of science-policy making in countries such as France, Germany, and Britain, a political space was constructed to deploy practices of biotechnology policy, with the state represented as a central force for reshaping the institutional complexes that described the science-industry interface, such as the relationship between the universities and industry. This space constituted an important semantic frame for the hegemonic struggle over the new biomolecular order.[3] Representational strategies advanced by scientists, administrators, and industrialists connected molecular biology with something very specific: a new understanding of life, and with it a dramatic increase and sophistication in the control of life processes. These processes could potentially be used in various ways in fields as diverse as the medical sciences or the "agrofood" industry and thus could contribute significantly to the solution of the major problems of humanity. In these textual strategies a new representational matrix was outlined connecting the discursive space of molecular biology with the delineation of a new economic space as the field of commercial expansion for the new biotechnology.

In order to understand the "logic" behind the British state's intervention in the economy, however, we have to look at some other crucial elements of the discursive economy of the 1970s that delineated the semiotic repertoires for constructing public policies on biotechnology. In particular, I want to point to the interconnection between theories of economic change and ideologies of neoliberalism which in a process of mutual reinforcement redefined the logic, scope, and content of public policies related to science and technology.

The economic discourse on innovation in these years was dominated by the "science push theory," the argument that the key to innovation was the "push" of scientific development.[4] And, according to a variety of studies conducted during this period, it was the state that had to "push" science. Especially during the 1960s and 1970s, most industrialized countries went about developing science and technology policies more systematically than in previous periods. The economic discourse generally identified science and technology as contributing to progress and modernization and associated it with economic growth, agricultural expansion, and medical advances. Scientific and technological progress was increasingly represented as a central condition for a nation's viability and ability to compete in the international economy. With these discursive changes a new interdiscursive economy emerged, a new field of concomitance, attributing elevated status to the growth and thriving of science.[5] Science and technology became increasingly defined as crucial to national pride, prestige, and viability.

The 1970s witnessed not only the arrival of recombinant DNA technologies and advanced technologies of cell fusion, but also significant socioeconomic transformation. The traditional meaning of "the political" seemed to change as citizens protested against nuclear power and mobilized in a wave of political activity from the women's movement to the peace movement.[6] It was in this political atmosphere that Paul Berg and ten other scientists sent their famous letter to *Science* on the potential biohazards of recombinant DNA molecules. With the oil crisis, surging inflation, and the economic recession in the first half of the 1970s, the dream of the never-ending post-World War II economic boom ended abruptly. European industry was hit especially hard by the economic crisis.

During the recession American semiconductor firms used the occasion to cut prices in order to gain market shares. Many European companies pulled out of the "commodity chip" markets and retreated into the relatively safer "custom chip" markets. The semiconductor market effectively became divided into two noncompeting sectors, the big league firms dominated by U.S. companies and the smaller firms in the custom chip segment. In the early 1980s it became clear that microelectronics and its applications in manufacturing and information technology were going to be vital for all industries. Not surprisingly, the competitive advantage of U.S. and, a bit later, Japanese firms in semiconductors, computers, telecommunications, and consumer electronics raised increasingly the painful question: Were European industries still viable in these crucial technological sectors?[7] During the 1980s the European Community and its member states responded with rescue operations for the European information technologies, without great success, however.[8]

In the early 1980s only true optimists believed that the European semiconductor, telecommunications, computer, consumer, and electronics industries could be effectively "saved." It was around this time that three central metaphors were increasingly used to construe the situation: Europe's viability was depicted as threatened, a technology gap between Japan, the United States, and Europe was diagnosed, and Europe was seen to be in a technology race with its two main competitors—Japan and the United States. This political discourse was readily supported by the interpretations of economists and innovation experts specializing in technological development, innovation, and trade policy. They expressed a consensus: Europe—that is, Western Europe—could only survive and thrive if it were to engage in the high-tech race and fight its technological and economic enemies Japan and the United States, by vigorously supporting and encouraging its national industries and research systems to innovate, cooperate, and compete. Success on the technological front became increasingly represented as a strategy to keep Europe what it used to be, to save its identity. This could be accomplished by fighting the others, be it the United States, inert national industries, or critics of technology. Europe ought to overcome internal divisions, whether of markets or of dissenting citizens, and unite to face the common challenge of the international marketplace.[9]

The representation of a Europe engaged in a "high-tech war" with Japan and the United States was part of a textual strategy, creating the political space that delineated the neoliberal reorientation that began in the United States and Europe around this time. The target of these strategies was the postwar capitalist growth model. That model was represented as containing a set of barriers, such as powerful trade unions, state monopolies, and lack of investment incentives, which together inhibited market dynamics and kept national economies and all of Europe from adapting and facing up to the new opportunities and challenges. Only by removing these barriers could recovery be expected. Elements of this "strategy of recovery" enacted in various policy fields included increasing the profitability of private capital, creating a more favorable labor market by dismantling the trade unions, reducing corporate and middle-class taxation, and privatizing public industries and banks.[10]

Science and technology had a strategic place in this project: European survival was read as only to be secured by facing up to the high-tech race, which in turn was possible only if structural impediments to innovation were removed. Hence, technology policies, trade policies, industrial policies, and social policies were positioned as mutually reinforcing discursive practices as the political order was reshaped, along with previously established notions of the interaction among state, society, and economy in post-World War II Europe.

In the context of Europe's industrial crisis, attention turned increasingly to what was widely perceived as the second important, broad field for high-tech innovation after information technology: the new biotechnology. The new biotechnology was expected to have its strongest impact on the chemical and pharmaceutical industry and on the agro-food sector. These sectors were the strongholds of European industry. European dominance in chemistry was especially obvious, with the German companies BASF, Bayer, and Hoechst as the world's largest chemical companies, followed by the British ICI.[11] In the early 1980s, however, as a result of a recession, the chemical industry was in decline and the European pharmaceutical industry, though outspending the U.S. companies in R&D, still lagged behind them overall.

Britain was faced by particularly harsh economic problems and industrial decline. While still under the Labour government, the new biotechnology became an agenda of national importance, and the state began to act on its behalf. In the late 1970s and early 1980s British biotechnology policies were shaped in several key documents, of which the 1980 Spinks report is the most important. Biotechnology policy became a surface of inscription interconnecting scientific struggles for disciplinary hegemony with political practices geared towards rewriting the post-World War II political order along neoliberal lines. Articulatory practices delineated a new spacing of the political by rewriting institutional complexes such as regulatory systems or company behavior through constructing new nodal points that fixed the meaning of the multifaceted process of biotechnology and its location in the sociopolitical matrix.

Part of this politics of meaning entailed inscribing genetic engineering as a technology whose benefits outweigh its risks. Only a few years earlier, in the mid-1970s, a broad and animated debate had taken place on the risks involved in genetic engineering. The new biotechnology inspired serious doubts and concerns. These debates eventually faded and led to the establishment of relatively unrestrictive regulatory measures. The practices of molecular biology had succeeded in defining the nature of the hazards associated with genetic engineering, and, as a consequence, they were seen as minor.[12]

In this respect, not only research policies but also the tightly intertwined regulatory policies became surfaces of inscription for scientific practices, extending their reach from the laboratory to society by means of genuinely political technologies such as laws and allocation of resources. Biotechnology policy both reflected a condensation of interscientific power struggles and contributed to the social construction of biotechnology by positioning it within the political discourse. The textual strategies deployed outlined a new representational matrix, a political space establishing a relationship between the discursive space of molecular biology and the newly emerging economic space of biotechnology.

Near the end of the Labour government in the late 1970s the contours of a British strategy toward biotechnology became visible. Britain's strategy for biotechnology had three major aims: (1) to delineate the problems to be solved by biotechnology and to construct a rationale for why biotechnology policy is important; (2) to define the field of biotechnology, to inscribe biotechnology as science policy; and (3) to suggest a strategy "to face up to the challenge," to rebuild existing organizational elements in order to enable biotechnology to develop. An important part of writing biotechnology policies was to create a rationale for the support and diffusion of biotechnology. As mentioned before, the crisis of capital accumulation and the political repercussions of the 1970s were in general the crucial rationale used to justify biotechnology policies.

The perception of a crisis was most strongly articulated in the viability metaphor, the interpretation that European industry lacked the dynamics and the ability to adapt to changed circumstances. In 1979, still during the British Labour government, the Advisory Council for Applied Research and Development (ACARD), the Advisory Board for the Research Councils, and the Royal Society set up a Joint Working Party (the Spinks Committee) to review existing and prospective science and technology relevant to industrial opportunities in biotechnology. Scientists, industrialists, and government officials increasingly felt that "something had to happen."[13] As one government official put it: "People felt that there was potential application of biotechnology which needed to be encouraged to a greater extent by using public funds. . . . It was a general feeling, we had microelectronics, the next is going to be biotechnology. " Another civil servant saw as the driving force behind the building of a British biotechnology strategy the developments in the United States: "I guess the impetus for it really came from all these investments in the small companies in the States and a feeling that there was a race we are losing in some way."[14]

The focus of the various governmental representations of the new biotechnology was on what was interpreted as its enormous potential to solve problems—to counter threats to health, the environment, agriculture, and the state of prosperity in general.[15] The United States and Japan enjoyed a ubiquitous presence in the various analyses. Although much of the evaluation revolved around first trends, potentials, and futures, documents constantly alluded to an already existing, or at least threatening, U.S. and Japanese superiority in biotechnology, a scenario perceived to undermine European industry and wealth. Modernization, economic growth, industrialization of agriculture, and industrial innovation enter this evaluation as unquestioned elements in need of science and technology for further development.

Another element of biotechnology policy was to map and outline the notion of "biotechnology."[16] For the state to become active, it was crucial to delineate commonly agreed upon dimensions, thought to constitute biotechnology, in order to isolate possible points of political intervention. As another official I interviewed put it, "Biotechnology does not really exist, but it is good to have a word for it."[17] The various actors tended to agree that "new" or "modern" biotechnology had something to do with the integrated use of biochemistry, microbiology, and engineering sciences as redefined by the use of genetic technologies. But this was as far as the tacit compromise went.

Broader and narrower definitions existed, but basically biotechnology was defined much as the state or other actors such as the research councils wanted it to be defined at a particular point in time. When biotechnology policies were being devised, the issue was a matter less of giving a definition than of having the power to define situations that would be read as biotechnological research. The term *biotechnology* operated more as a semiotic, polysemic resource that could be used more or less successfully in strategic situations, and not as a relatively clearly defined term like, for example, genetic engineering.

An example of the strategic use of the term *biotechnology* can be found in the British debate on the issue. Of all the research councils, the Medical Research Council (MRC) most fiercely resisted attempts by the Department of Trade and Industry (DTI) to define its research priorities. Defining specific fields of research was part of the government's strategic approach: its definition of biotechnological research was more oriented toward potential industrial applications than that of the MRC. To counter the assertion that it was out of line with the government's plans, and to avoid the DTI's imposing its interpretation of the field, the MRC resorted to the strategy of arguing that it had already been engaged in biotechnological research for decades—in essence redefining its history of biomedical research as the history of biotechnology.[18]

The definition of biotechnology strategies proceeded within carefully crafted semantic fields depicting the importance and effectiveness of the new biotechnology. The strategies focused on removing of "biotechnology obstacles" and launching "biotechnology offensives." Major obstacles to a successful biotechnology strategy were located mainly in the way that research and industry were organized, as well as in a lack of "public understanding." The Spinks Committee report came to the following devastating analysis of the state of biotechnology in Britain:

> The present structure of public and private support for R and D is not well suited to the development of a subject like biotechnology which, at the moment, straddles the divisions of responsibility both among Government Departments and Research Councils and the arbitrarily defined fields of fundamental and applied research. Strategic applied research is in general ill-served by our research funding mechanisms, especially in areas where there are neither university departments to promote it, nor well-developed industries to provide market-pull. Biotechnology lacks university centers specifically for it and this results in a shortage of new ideas for industry and of suitable trained manpower. The absence of an established industry clearly identified with new biotechnologies allows Government policy in one industrial sector to have adverse implications for biotechnology in others and diminishes the attraction of the subject for the engineers. The result of such interactions is that biotechnology has not grown in the United Kingdom in the coherent fashion which we believe is merited by its potential. What is required at this state is a policy of "technology-push" reflected in a firm commitment to a strategic applied research. This will progressively produce potentially marketable products and processes and the policy should then be for a more "market pull" approach.[19]

What this analysis suggests is that Britain had neither a biotech industry nor a research system conducive to it, and, hence that the national innovation system had to be rebuilt. A strategy was required. The key elements in this strategy were sufficient funds, strong engagement of the research councils through substantially increased support and coordination of activities in a Joint Committee for Biotechnology, and coordination of the government departments and creation of an interdepartmental steering group. This strategy is depicted as crucial in the face of worldwide competition. Currently, the report argues implicitly, there are two models for developing a biotechnology industry: the U.S. model, with little government support but strong industrial research and a sprawling venture capital industry; and the European-Japanese model, with heavy government involvement.[20]

The Spinks report was a first effort to enframe the meaning of a British biotechnology policy and to create a regime of signification defining the interests and coordinating and stabilizing the practices of a variety of actor positions configured in institutional complexes such as government or firms. But the Spinks report, like any other government report, succeeded as a framing device only insofar as it regulated the balance between dissensus and consensus of the actors. I argue, in other words, that the success of science and technology policy depends on organizational conditions allowing the state to reduce the "messiness" of the political process, to mitigate political conflicts, and to foster cooperation by creating a political space that permits the articulation and coexistence of differences in the process of policy making. Such a political space is an essential prerequisite for the transfer of laboratory practices into the "outside" world, in other words, for the production of the "new economic space of biotechnology."

Multiple Readings of Biotechnology Strategies Under the Thatcher Government

The Spinks report was an anachronistic document soon after it was published. With the 1979 elections the rhetoric of the "high-tech biotech race" gained a new meaning in the political discourse of Thatcherism.

The political character of the postwar period in Britain was defined by the "settlement" of the 1940s. An unwritten social contract had emerged, with the Right accepting the welfare state, comprehensive education, and Keynsian management of the economy, as well as a commitment to full employment as the terms of a peaceful settlement with labor, and with the Left settling for working within the framework of a modified capitalism and within the Western bloc's sphere of influence. After the "restoration" of capitalist imperatives during the 1950s, within the framework of U.S. world hegemony, the 1960s and 1970s were dominated by the Labour party. Labour attempted to manage the new big state-big capital corporatist arrangements developed as the basis of economic policy and planning, by harnessing the working classes to the corporatist bargains through the trade unions and disciplining them through Labour's traditional alliance with the unions.

But the underlying economic conditions for this stabilization, in particular as articulated by the Wilson government, did not exist. The British economy was too weak and undercapitalized to sustain capital accumulation and profitability and to continue to subsidize the social compromise made after the 1940s. As the world underwent economic recession in the 1970s, British society began to polarize under conflicting pressures, and the social compromise eroded. At this time the New Conservatism emerged under Margaret Thatcher.[21]

The strategic goal of Thatcherism was to embrace a distinctively neoliberal, free-market system and to break the power of the working class, but also to curb industry's influence on the state. At the same time this free-market approach was linked to traditional issues of Toryism: a return to the old values, Englishness, family, and nation. Despite its official rhetoric of commitment to the virtues of market-led solutions to economic problems, the Conservative governments since 1979 have in fact presided over a renaissance of intervention. These governments did shift from a "relatively extensive" to a "relatively intensive" regime of intervention. Whereas in the past the state intervened in the economy through subsidization, economic regulation, and nationalization, the Thatcher government shifted to a new strategy. First, the field in which intervention operated was narrowed in scope and extent, but this intervention also became more specific. Second, economic intervention was deepened and made more selective, discriminatory, and value-for-money oriented. This heightened selectivity went along with better scrutiny, such as better forms of project assessment and follow-up studies on finished projects. Third, a new layered system of intervention was created, a policy to create overlapping mechanisms of regulation. Fourth, greater discretion was granted to the units of intervention, giving them greater autonomy and space to maneuver within the system.[22]

What was at the core of the economic strategy of Thatcherism was not less state, but a "redifferentiation" of state intervention, which by no means meant simply "less state intervention," or a state intervention geared primarily towards the interests of capital, but a new mode of regulation. On the contrary, Thatcherism put a strong emphasis on "liberating" the state from the influence of both labor and capital as institutionalized in corporatist bargaining.

With respect to labor this strategy translated into a systematic effort to remove, weaken, or radically reform those institutions that had promoted unionism and, in the view of the government, had adversely affected labor costs, productivity, and jobs.[23] With respect to industry the Thatcher government continued a tradition of government-industry relations inherited from the nineteenth century: at the core of this approach was the assumption that preserving the commercial autonomy of firms is of primary importance, and that an arm's-length relationship between the government and industry is generally acceptable.[24] Furthermore, much as with labor, the Confederation of British Industry (CBI) saw its influence on governmental decision making erode: in fact, ministers outmaneuvered and disarmed the CBI systematically. However, the new strategy of regulation did not mean that all previous modes of intervention were abandoned. Subsidies, for example, continue for obvious reasons to occupy a prominent role in the repertoire of regulatory mechanisms.

But like other aspects of the new course of neoliberal policy making, such as privatization, the new, "relatively intensive" strategy had to prove its value in the everyday life of political and economic behavior. Reorganization on the level of the main macroeconomic front line of politics and a new politics of representing state-society interaction does not necessarily translate into the microeconomics of managerial commitment and investment.[25]

Last, liberating the state from the obligation of neocorporatist bargaining—a move based on the assumption that the state would gain autonomy as a result—was a single tactic but an important one in the project to construct new spheres of state autonomy. The neoliberal project included articulatory practices that resignified the autonomy of different units of the innovation system—regulating relations between, for instance, DTI and the Science and Engineering Research Council (SERC) to allow the new strategy of relatively intensive intervention to be implemented. Objectives such as better scrutiny of supported projects or increased discretion for intervening mechanisms, for example, the Biotechnology Unit of the DTI, required an absence of antagonism within the innovation system in order to establish "smooth" entry points for intended interventions, and this implied constructing mechanisms for coordination.

Soon after the ascent of the Conservatives to power it became clear that the new neoliberal political discourse created a new meaning for biotechnology policy. The strategy developed for biotechnology followed closely the pattern of "relatively intensive" intervention. The first major government statement on biotechnology after the change in power followed many of the key recommendations of the Spinks report, but indicated a change in the strategy for pursuing the outlined goals. The government declared its belief "that biotechnology will be of key importance in the world economy in the next century, and of rapidly growing importance before then." The white paper made the following central recommendations: (1) To strengthen the scientific base of biotechnology within the framework of the existing framework of funding, mainly by concentrating, implicitly meaning, shifting resources. (2) To coordinate the government's biotechnology-related activities between government departments and between those departments, the research councils, the National Research Development Corporation, the National Economic Development Council, and the National Enterprise Board. (3) To coordinate and foster collaboration between universities, research councils, and industry. (4) To encourage the increase of private investment in the biotechnology industry. (5) To remove regulatory constraints inhibiting biotechnological development, such as certain health and safety regulations. (6) To foster international collaboration and competition.[26]

With this strategy the government had clearly moved away from the much more interventionist "Spinks philosophy," but at the same time it had outlined a project of intergovernmental coordination, closely linked to the plan to coordinate the activities of the research councils, to press industry not to underinvest but to internationalize research and production. This targeted strategy focused less on increasing government resources for biotechnology than on a more efficient use of existing resources and facilities, on the national as well as on the international level—a strategy which, however, required substantial intervention from the state and, to be sure, the cooperation of industry.

The Thatcher government's new framing of biotechnology policy constituted an effort to redefine the interests of the "illusory community" of those it claimed to represent. During this spacing of the state-society interaction, those nodal points interconnecting the national innovation system were being reconstructed, thus reconfiguring biotechnology policy from a mode of extensive to a mode of intensive intervention. The nodal points in a chain of significations describing the new meaning that the government attributed to biotechnology policy comprised generating a "new climate for investments," improving cooperation between the academic R & D system and industry, establishing new modes for subsidizing research in public and private institutions, and a politics of "enforcing" higher R & D expenditures by industry.

The power change in 1979 did not put an end to discussion on a British strategy for biotechnology. By then a pattern of dissensus on the future of British biotechnology policy had already been established, reflecting irreconcilable differences in the reading of biotechnology by a variety of key actors. In a 1981 meeting of the British Coordinating Committee for Biotechnology drawn from various learned societies and professional institutions at the Society for Chemical Industry in London, representatives from government, science, and industry made a bleak assessment of the future of biotechnology in Britain, citing in particular the "failure" of the government to respond to the Spinks report and to take measures like those taken in Japan and Germany. The participants frequently pointed to the lack of coordination between the research councils, industry, and research and the difficulties in financing innovation related to biotechnology. As one participant noted, "It is not possible to fund research with venture capital because the time scale is wrong. Venture capital is attracted only to high growth areas and, while biotechnology is new and exciting, it has not shown any growth yet. Further, it tends to fall within existing industrial areas which had been hit by the recession. The real inventions . . . are to be found within universities, but these will not attract investment unless academia identify markets as well as make inventions."[27]

In 1982 the Education, Science and Arts Committee of the House of Commons came out with a report on the issues subtitled "Interim Report on the Protection of the Research Base in Biotechnology." This document again criticized the lack of government initiative, the lack of coordination between the institutions responsible for biotechnology policy, the cutbacks in the science budgets, and industry's pattern of underinvestment.[28]

In October 1982 the Trade Union Congress (TUC) General Council approved a paper asking for a comprehensive and planned approach toward biotechnology policy to do justice to what TUC saw as the potentially huge benefits to the economy and to society as a whole. TUC essentially declared its support for the Spinks strategy of broad-scale intervention, though mentioning the need to account for social implications.[29] But neither the unions nor industry were capable of exerting a visible impact on the government's evolving strategy for biotechnology in the United Kingdom. Government biotechnology strategy took a new tack when Patrick Jenkins replaced Keith Joseph, one of the key architects of Thatcherism and a free-market advocate, at the top of the Department of Industry (DoI) in 1981. This move took place as early hopes vested in small businesses as centers of enterprise and creators of jobs began to fail. The new minister generally tried to carve out a more active role for his department in tune with its pre-Thatcherian traditions.[30]

As one of the main architects of British biotechnology policy in the DoI remembers:

Patrick Jenkins was interested in biological applications. He had worked for distilleries and with yeast. He had been involved with biological research and chemical research: They said he wanted to have something done, . . . they came to me and said: "Help! The Secretary of State wants us to have an initiative in biotechnology. This was in February 1982. They asked: "What can you do in a month?" And I said: "Nothing. I said I'll give you something by the autumn. . . . We were talking to industry [about] what it is we should do, not on the basic research, but in the way we can help industry to pick up biotechnology. We also had an outside consulting group looking at biotechnology. We went around between March and August and talked to everybody we could lay our hands on in industry and academia [asking] what they thought, what should be done, and then, in September we wrote our report and we said, yes, there is a role for government for these various things which should be done and please can we have 16 million pounds over three years.[31]

By the end of 1982 the DoI was the lead department in biotechnology, with overall responsibility for encouraging industrial developments and for those areas that did not fall clearly into the provinces of Department of Health and Social Security, the Department of Energy, and the Ministry of Agriculture, Fisheries and Food. An Interdepartmental Committee on Biotechnology had been established to coordinate the interests of government departments and related bodies. On the level of the research base the Department of Education and Science made extra funds available to University Grants Committee, which allotted special funding for biotechnology. The Agriculture and Food Research Council (AFRC) had diverted staff from lower-priority work to reinforce its biotechnology programs; the MRC increased its support for biotechnology; and the SERC had established a directorate with funds earmarked for biotechnology, with total grants expected to rise from £1.6 million in 1982–83 to £3.5 million in 1985–86. The SERC identified certain key sectors of biotechnology and proposed to establish centers of research expertise in them. The DoI set up a new Biotechnology Unit at the laboratory of the Government Chemist to promote its long-term biotechnology program in November 1982, with a budget of £15 million.[32]

The industry secretary described these developments forthrightly:

Government is committed to sustained support for the development of biotechnology in the UK. The diversity of the technology means that a battery of measures is needed. Current Government support for biotechnology (30–40 M per annum depending on strictness of definition) is comparable with that in France, Germany and Japan. There is no room for complacency, but Government is determined to help industry develop new and profitable products and processes arising from our excellent research base in biotechnology.[33]

Clearly, the Conservative government was determined to intervene in the shaping of biotechnology in Great Britain. Yet despite strong social pressures from science, industry, and labor for a more comprehensive intervention of the state in biotechnology, the government insisted on its "relatively intensive" strategy. But this strategy failed to achieve the envisioned impact.

Strong governmental commitment to promoting biotechnology was subverted by the particular strategy adopted to accomplish that promotion. This strategy was based on a spacing of the national system of innovation that was soon derailed by an emerging pattern of antagonism and contradiction, effectively blocking significant parts of the government's policy schemes.

The Thatcher Strategy Subverted: The Economic Context

There is clear evidence that the Thatcher government could formulate a biotechnology policy based on a compromise between industry, science, and the state, and that state support for biotechnology was generally considered desirable. But government biotechnology policy found itself constricted by the macroeconomic context it was embedded in. The Thatcher cabinet's politics of neoliberalism—its economic policy, industrial policy, and politics of industrial relations—failed to break the long-term trend of economic decline. The dramatic fall in new capital formation between 1979 and 1985, modest levels of investment, the inferior record of investment in labor-force training, and the poor record of nonmilitary industrial research were indicators of this failure.[34]

This macroeconomic context certainly helps explain why small and medium-sized companies played only a minor role in opening up the economic space of biotechnology. These economic difficulties also explain in part why the large chemical and pharmaceutical companies did not compensate for this lack of small companies (see below), as their initial reluctance to engage in biotechnology-related investments was reinforced by knowledge that many U.S. biotechnology companies were still not operating at a profit. (However, companies like ICI, Glaxo, and Wellcome have considerably increased their investments in biotechnology over the last ten years.[35])

Furthermore, the Thatcher government's macroeconomic strategy put a priority on deficit reduction and on anti-inflationary measures that reduced government resources. These strategies had a direct impact on its science and technology policy and on economic demand. While the British government was arguing that there was no room for a Keynsian-style expansion of the domestic economy, the financial flows were channeled from the United Kingdom into U.S. government debt and helped stimulate the U.S. economy instead.[36]

This economic policy had a number of negative impacts on biotechnology R & D. Cuts in the science and technology budgets affected the universities and the AFRC institutes most severely, in particular from the 1990s on. The decline in the infrastructure of the universities reduced their operational budgets, and increased teaching responsibilities reduced the time faculty members spent in research. The AFRC was forced to close most of its institutes and eliminate whole categories of research in the remaining ones. At the same time the 1980s saw an overall expansion of biotechnology-related R & D expenditure (which did not necessarily help the AFRC or all the university departments). The public-sector funding expanded from £58.9 million in 1985–86, to £68.7 million in 1987–88, to £146 million in 1990, and industrial spending from £48.6 million in 1988, to £58.1 million in 1989, to £73.5 million in 1990.[37] Hence, the general cuts in the British R & D budgets were more than offset by specific increases in governmental and industrial support for biotechnology-related research. Money, then, was not the problem; rather, it was the government's biotechnology strategy and the organizational features of the British innovation system.

The practices of British financial capital account for many of the difficulties of creating an economic space for biotechnology in Britain. Influenced by the U.S. model of small companies opening up the economic space of biotechnology, the government made financing these small companies a crucial issue in its biotechnology strategy. But within the framework of the Thatcherite discourse on economic policy, the only language available for dealing with the issue of the relationship between financial and industrial capital was that of deregulation. This policy exacerbated a long-lasting British tradition.

Like the United States, Britain historically relied on private capital to finance industry. As Alexander Gerschenkron has pointed out, Britain did not develop a system of long-term finance to promote industrial growth because none was needed in this first and unique case of industrialization.[38] Unlike the United States, Britain does have a history of government interference in industrial policy. Whereas the government treated the financial system as sacrosanct, it intervened constantly and often as an adversary in industry. The result was a contradictory financial and industrial policy, to a considerable extent responsible for Britain's deteriorating economic growth, compared with most of its competitors; increased decline in the share of world trade; inflationary instability; rising unemployment; declining living standards; and extremely low formation of capital.[39]

No postwar British government ever intervened directly to control credit and change. This attachment to traditional private capital market practices created fundamental obstacles to financing industry. Not only was there no tradition of industrial banking and short-term lending, but the stock exchange itself was never an effective source of lending.

This anti-industrial trend was reinforced when the City of London became largely an offshore center in response to the expansion of the Eurodollar market. The private capital markets, left to their own devices, pursued policies that led to money being invested more in property, overseas industry, and government debt than in domestic industry. The Thatcher government fit its policies into the existing structure of the financial system, consolidating Britain's specialized role as the principal site for international foreign institutions by abolishing exchange controls, deregulating financial institutions, and granting favorable tax treatment. This policy created a financial system geared more to overseas investment, with no tradition of industrial banking, an underfunded and uncompetitive industrial base, and an adversarial political system incapable of arriving at national consensus over economic policy.[40]

Furthermore, the emerging venture capital industry, construed as a fruit of deregulation, failed to fulfill its promise as a source of capital for new risk investments. Although the venture capital industry grew markedly after the 1980s, the United Kingdom, in dramatic contrast to the United States, saw its venture capital firms allocate relatively little investment to technology-related, start-up, and early stage deals. In contrast to the early days of the 1990s, these funds are more likely to put money in management buyouts and company development than in start-ups. And even the growth of the British venture capital industry must be put in perspective: In 1991, for example, the whole of European venture capital funds of 88 million ecus did not even match the $114 million that Genentech spent on R & D in the same year.[41] Such practices of British financial capital vis-à-vis industrial production proved a major obstacle for biotechnology investments.

When the economic space of biotechnology first opened up in Britain, large companies such as Wellcome and ICI were reluctant to invest in it. Probably more troublesome for the government's declared goal of attracting small and middle-sized companies to the new biotech industry were organizational features of the financial markets that effectively raised major obstacles to the creation and survival of new and smaller companies. Since biotechnology policy was contextualized within the larger framework of a politics of further deregulation of financial markets, the government did little to reinforce its intervention in the biotechnology industry by resorting to concrete intervention in the finance industry's relation to biotech firms.

Deregulating and leaving the financial markets to themselves did not, however, make financial institutions less rigid. Among the rigidities of the British financial markets are the requirements for listing a company on the stock exchange, a crucial device for companies to raise money. Whereas in the United States the National Association of Securities Dealers exchange for small stocks (NASDAQ) has few rules, the London Stock Exchange (LSE) requires companies to show revenue, profits, and minimum trading experience before they can be listed. As a result, the only biotechnology company quoted at the LSE is the British Bio-technology Group in Oxford. In 1992 the first British biotechnology company, Cantab, opted for a NASDAQ listing. Concerned that other companies would follow Cantab's lead, the LSE began to reconsider its policies. Under the new regulations, however, only those companies producing human pharmaceuticals need apply. As an LSE official put it: " We believe that only biotechnology companies developing human drugs in clinical trials can provide the level of comfort we desire." Companies operating in sectors like agriculture, diagnostics, and manufacturing are be excluded. Furthermore, pharmaceutical companies need to demonstrate their R & D competence with patent applications or awards and have at least two drugs developed in-house being tested in clinical research.[42]

Not surprisingly, a recent survey shows fewer than fifty new biotechnology companies operating in Britain, the vast majority of which have just a handful of employees and flat growth patterns predicted. Only six of these companies appear to have the ambition and potential to become significant, multiproduct companies.[43] Unable to rely on an initial public offering (IPO), the typical and profitable route of divestiture for venture capitalists in the United States, venture capitalists in the United Kingdom refrained from investing in biotechnology and concentrated on other areas of investment, such as computers.

Within the general context of the comparatively rigid British financial system, in consequence, various government schemes to promote collaboration between science and industry could have only limited impact. Whereas government programs did attract large companies, the companies considered to be especially valuable, the small and medium-sized companies, could not engage in such collaborations, mostly because of a shortage of funds.[44]

The organization of the financial markets and their further deregulation generated practices that considerably destabilized the state biotechnology strategy. That policy made the financial system a nodal point in its writing of the British biotechnology industry, but the financial markets simply did not generate the capital necessary for investments in that area. What were missing were political practices, such as a negotiated regulatory structure, that could make productive venture-capital investments more attractive. This would have required intervention in the financial markets, a move ruled out for ideological reasons.

The Research Councils' Reading of Biotechnology Policy

Implementing the government's biotechnology strategy, however, depended not only on the cooperation of financial capital, but also on how industry and the R & D system were organized. The government's biotechnology policy had as explicit objectives reorganizing the public R & D system to use resources more efficiently, coordinate better the various units of the R & D system, and improve collaboration between public research and industry. But government could not dictate these objectives or other goals, such as the better scrutiny of supported projects. To do so required establishing organizational mechanisms to handle antagonism between the various units of the biotechnology innovation system and to facilitate cooperation.

Within the framework of British science and technology policy, basic research is funded through the University Grants Committee (UGC) and the research councils, both under the Department of Education and Science. Whereas the UGC gives block grants to each university to cover both educational and research expenses, the research councils support more targeted research, focusing on particular areas such as agriculture and medicine. Industrial innovation is the responsibility of the Department of Trade and Industry (DTI)[45] through its Biotechnology Unit and the British Technology Group.

The existence of four research councils with partially overlapping responsibilities and a perceived need for coordination created one of the central challenges to constructing a British biotechnology policy. The DTI's Biotechnology Unit, in particular, which was responsible for developing and promoting industrial activities in this sector, made consistent efforts to improve the coordination between itself and the research councils.

Following a recommendation of the Spinks report, an Inter-Research Council Coordinating Committee for Biotechnology was established—and soon dissolved as largely inefficient. It was succeeded by the Biotechnology Advisory Group, which itself was succeeded in 1989 by the Biotechnology Joint Advisory Board established by the SERC and the DTI, which were joined by the AFRC, the MRC, and the National Environmental Research Council (NERC) in 1990. The attempts to find a suitable organizational structure for funding decisions in biotechnology hence lasted more than ten years; in the light of the current policy debates in the United Kingdom, it still seems to be unresolved.

The government's biotechnology policy was in part to be written by the research councils, which controlled vast resources and the discretionary power to channel the resources. As a DTI representative put it:

> There is a strong tradition of hands off from the research funding agencies, there is no way to instruct them what to do, and there is a very strong and widely accepted view that there should be academic freedom and the Research Councils are fiercely independent organizations and do not think it would have been efficient if some minister said, "Look here, chaps, this is what you should do, and that is not the way that any impact would be made."[46]

Historically the research councils' autonomy dates to the reorganization of British science after World War I. In 1918 a committee of inquiry into the machinery of British government was set up under Lord Haldane. Its report argued that research done for broader purposes than for a particular government department should not be supervised by any particular department or government institution. What was later called the Haldane principle of research council autonomy supplied important justification for the aforementioned "hands off" rhetoric.[47]

But the Haldane principle in fact grew out of a longer British tradition of disjunction between science and its application. Whereas nations with smaller colonial markets and delayed industrialization created scientific organizations closely tied to industrial process, Britain, with its successful industrialization, was under less pressure to develop a similar model.[48] Thus developed a tradition and pattern of state nonintervention in some of the key institutions of the innovation process, most notably the financial markets and the research system.

The other result of this tradition was that antagonistic relationships quickly emerged once the government tried to discontinue the pattern of noninterference. Efforts to improve coordination between the councils were frequently interpreted as efforts to eliminate identities: the MRC in particular saw its identity and history as one of the proudest research institutions in Great Britain at risk in this way. Under these conditions creating systems for cooperation turned out to be difficult.

The MRC and the AFRC proved strong enough to resist persistent government efforts to bind them in a common decision making structure headed by the DTI. When these councils turned to a more application-oriented approach, as when the MRC cooperated with Celltech, or the AFRC with the Agricultural Genetic Company (AGC) (established in 1983), they were driven more by the drying up of government funds than by consent or a new corporate philosophy. Moreover, other research council actions betray a similar lack of coordination with the government's general biotechnology policy. The MRC's relationship with Celltech (1980–83) and the AFRC's with AGC were initially exclusive. This structure kept other companies from cooperating with these two research councils and hindered them from transferring technology more broadly from the R & D system to industry.

The result was that research activities of the MRC, SERC, and AFRC overlapped significantly in basic molecular biology and genetics, as well as in some areas of biochemistry, biophysics, and cell biology. After the SERC established its Biotechnology Directorate, its relationships with the MRC and the AFRC deteriorated, since these councils felt that the SERC targeted research areas that fell in their province. Furthermore, the technology transfer policies of the MRC and the AFRC conflicted with the government's strategy.

Nonetheless, the existence of separate research councils with different understandings of the nature of biotechnology research did not per se destabilize the government biotechnology policy. Certainly, there were significant cultural differences between the research councils. The SERC, for example, followed a problem-solving engineering approach to defining objectives for biotechnology research and structured the research accordingly, whereas the MRC rejected any goal-oriented direction of research. Taking into account the very nature of biotechnological research—with increased blurring of the lines of demarcation between "applied" and "basic," "science" and "technology"—both councils could develop an interest in the same research problem, such as protein engineering, which could answer important questions about the nature of both proteins and drug design.[49] Such a situation does not necessarily imply conflict. On the contrary, one might argue that different approaches to protein design would lead to complementary research work. However, to facilitate cooperation when differences arose required an organizational structure that permitted negotiation. Without a government strategy to produce such a political space for negotiation, the established organizational logic maintained a pattern of antagonism and uncoordinated research between the research councils.

As a 1989 review of the research councils put it:

> Councils (AFRC, NERC, MRC) were often able to collaborate more effectively with each other than with SERC. Conflict has sometimes arisen from attempts to coordinate basic and strategic research activities promoted from the different perspectives of the mission-oriented Councils and of SERC. These conflicts reflect the lack of agreement on the extent to which SERC underpins the work of the mission-oriented Research Councils, an apparent lack of awareness by SERC of the basic research supported by the other Councils, and problems arising from different methods of working.[50]

It was not until 1993, when John Major took over from Thatcher and British science policy was generally reoriented, that a new research council was founded: the Biotechnology and Biological Research Council, now responsible for research ranging from basic biology to its industrial applications.[51] However, since this reorganization did not affect the MRC, how well it will solve the problem of coordinating the research councils remains an open question.

Besides coordination of the research councils another persistent problem, one only partially solved by government strategy, is that of transferring knowledge from the academic R & D system to industry.

A report on the SERC Biotechnology Directorate's relations with a sample of small and medium-sized firms gives clear evidence for that pattern. All of these firms conducted significant in-house research and made extensive use of academic research. The study identified shortage of funds as the main hindrance to academia-industry interaction. Besides this, governmental schemes apparently neither had found their target audience nor seemed to be very convincing: "Very few companies were able to distinguish the Biotechnology Directorate and its schemes from other organization's schemes. Nor did they have any clear idea of the technology transfer opportunities offered by the Biotechnology Directorate. In general, the companies thought that most of the schemes cost too much and covered too long a period."[52] Overall, it appears that the culture of university-industry interaction was only minimally developed:

> The most generally expressed industrial view about academic scientists was that their main role was to do basic research; . . . along the same lines was the view that universities' main task was to provide the "cannon fodder" for industrial recruitment, not to train new scientists. . . . For their part, many university researchers would prefer not to have to take industrial contacts at all. . . . Reliance on industrial funding, they fear, may deflect them from doing basic research.[53]

Clearly, these quotations reflect strong tensions between science and industry, a difference in organizational cultures, which impeded the process of technology transfer. Most strikingly, however, is the state's apparent failure to organize a scheme of technology transfer that would more successfully entice industry into collaborating with the DTI and the SERC. Government schemes continued to be drafted, but too often they failed to reach their clientele and to integrate it into the state's biotechnology policy.

Conclusions

This essay has focused on one of the key problems of policy analysis: the organizational difficulties that arise from the state's attempts to direct and shape institutions of civil society, such as industry and science. When, in the early 1980s, the British state began to intervene more systematically in building the "new economic space" of biotechnology, this case of innovation politics was represented as a crucial step for the British economy in an international "high-tech race."

However, more than ten years later, after a myriad conflicts and battles over "British biotechnology," the impact of the policies chosen remains highly uncertain, and some of the key initiatives clearly failed to achieve their objectives.

I emphasized in the beginning of the article how dramatic shifts in the discourse of molecular biology provided a newly shaped system of signification for the cognitive organization of British biotechnology policy. Within this discourse new representations of life as offered by molecular biology came to be associated with broad benefits for economy and society, a tendency accelerated by the development of recombinant DNA technologies in the 1970s. With the political-economic discourse stressing the connection between scientific and technological progress and economic growth and a need for a remaking of the post-World War II political order, a new interdiscursive economy emerged that elevated science and technology to a status of central national importance. The politics of molecular biology and, later, the politics of biotechnology must be located within this interdiscursive field.

This rhetoric that described international technology as a race and ascribed sociopolitical importance to biotechnology gained new meaning when power shifted from Labour to the Conservatives in 1979. The Thatcher regime's commitment to biotechnology was subverted by the very strategy adopted to foster biotechnology. Here we see ideology as describing positions in the political competition, tied to the organization of the political system, both playing an important role in shaping British biotechnology policy. Within the context of a political strategy indebted to a tradition that the British state did not intervene in the relationship between industry and finance, and the cultural strength of a "hands-off" tradition vis-à-vis the research councils, the organizational features of the British innovation system constituted a major obstacle to opening the economic space of biotechnology.

In this sense, the British state turned out to be strong and weak at the same time: strong in its efforts to regain autonomy from social forces such as labor and capital, weak in its capability to intervene. Missing were both an encompassing strategy of intervention and entry points for the intended intervention. The combination of overlapping research conducted by the research councils and the ensuing inefficient usage of funds, a general reduction of research money, and rigidities in the financial markets created a paradoxical gap between government rhetoric on biotechnology as future technology and government policy, which failed to address some of the central prerequisites for diffusion of biotechnology R & D into the economy. The ensuing interactions between industry, finance, and research made it difficult to use increasingly scarce resources efficiently and created serious obstacles to the transfer of knowledge from public research to a multitude of companies. The state's biotechnology strategy reflected the very same "messiness" it intended to manage.

Without a political space in which to control proliferating antagonisms and to facilitate cooperation, the British innovation system for biotechnology led to battles over turf and the dominance of a few large companies, rather than to a diversified biotechnology industry.

Notes

1. See, e.g., Sheldon Krimsky, *Genetic Alchemy: The Social History of the Recombinant DNA Controversy* (Cambridge, Mass.: MIT Press, 1982); and Susan Wright, *Molecular Politics: Developing American and British Regulatory Policy for Genetic Engineering, 1971–1982* (Chicago: University of Chicago Press, 1994).

2. Pnina G. Abir-Am, "From Multidisciplinary Collaboration to Transnational Objectivity: International Space as Constitutive of Molecular Biology, 1930–1970," in *Denationalizing Science: The Contexts of International Scientific Practice,* ed. Elisabeth Crawford, Terry Shinn, and Sverker Soerlin (Dordrecht: Kluwer Academic, 1993), pp. 153–86.

3. Pnina G. Abir-Am, "The Politics of Macromolecules: Molecular Biologists, Biochemists, and Rhetoric," *Osiris* 7 (1992), 164–92, on p. 166.

4. Rod Coombs, Paolo Saviotti, and Vivien Walsh, *Economics and Technological Change* (Totowa, N.J.: Rowman and Littlefield, 1987), pp. 223–29.

5. For overviews of the evolution of science and technology policy see J. Ronayne, *Science in Government* (Baltimore: Arnold, 1984); and Ros Herman, *The European Scientific Community* (Harlow, Essex: Longman, 1986). For field of concomitance see Michel Foucault, *The Archaeology of Knowledge and the Discourse on Language,* trans. A. M. Sheridan Smith (New York: Pantheon, 1972), pp. 57–58.

6. For important contributions on the impact of new social movements on society and politics see Claus Offe, "New Social Movements: Challenging the Boundaries of Institutional Politics," *Social Research* 52 (1985), 817–68; and Alberto Melucci, *Nomads of the Present: Social Movements and Individual Needs in Contemporary Society* (London: Hutchinson Radius, 1989).

7. Margaret Sharp, *European Technology: Does 1992 Matter?* Papers in Science, Technology, and Public Policy 19 (Essex: Science Policy Research Unit, University of Essex, 1989), pp. 4–6.

8. Philippe de Woot, *High Technology Europe: Strategic Issues for Global Competitiveness* (Oxford: Basil Blackwell, 1990).

9. The political construction of the global high-tech race is relatively neglected. The rhetoric of the "technological race" theory can be found in innumerable documents and public statements. For a good example delivered by a top European Commission official see Karl Heinz Narjes, "Europe's Technological Challenge: A View from the European Commission," *Science and Public Policy* 15 (1988), 395–402.

10. Patrick Camiller, "Beyond 1992: The Left and Europe," *New Left Review* 175 (1989), 5–17, on p. 7.

11. Luigi Orsenigo, *The Emergence of Biotechnology: Institutions and Markets in Industrial Innovation* (New York: St. Martin's Press, 1989); and Club de Bruxelles, "Bio-Industry in Europe" (MS, Brussels, 1990).

12. For a discussion of this process see Susan Wright, "The Society Warp of Science: Writing the History of Genetic Engineering Policy," *Science, Technology, and Human Values* 18 (1993), 79–101. See also interviews, Department of Trade and Industry, London, 9 July 1992, 8 July 1992. These interviews with various officials were granted on the promise of anonymity. To preserve that anonymity yet allow the reader to contextualize the source and content of my interview data, I will identify the interviews only by date, location, and the institutional affiliation of the interviewee.

13. Edward J. Yoxen, "Assessing Progress with Biotechnology," in *Science and Technology Policy in the 1980s and Beyond,* ed. Michael Gibbons, Philip Gummett, and Bhalchandra Udgankar (London: Longman, 1984), p. 223 n. 24.

14. Advisory Council for Applied Research and Development, Advisory Board of the Research Councils, and the Royal Society, *Biotechnology: Report of a Joint Working Party, March 1980* (London: HMSO, 1980).

15. Robert Bud, *The Uses of Life: A History of Biotechnology* (Cambridge: Cambridge University Press, 1993), p. 206.

16. For a rich microstudy on the definition of biotechnology in Quebec, see Alberto Cambrosio, Camille Limoges, and Denyse Pronovost, "Representing Biotechnology: An Ethnography of Quebec Science Policy," *Social Studies of Science* 20 (1990), 195–227.

17. Interview, Commission of the European Community, General Directorate XII, Brussels, 1 July 1992.

18. Interview, Department of Trade and Industry, London, 8 July 1992; and interview, Agricultural and Food Research Council, Swindon, 8 July 1992.

19. Advisory Council, *Biotechnology* (note 14), p. 8.

20. Advisory Council, *Biotechnology* (note 14), pp. 18–22.

21. Henk Overbeek, *Global Capitalism and National Decline: The Thatcher Decade in Perspective* (London: Unwin Hyman, 1990).

22. Graham Thompson, *The Political Economy of the New Right* (Boston: Twayne, 1990), pp. 135–42.

23. Steve Evans, Keith Ewing, and Peter Nolan, "Industrial Relations and the British Economy in the 1990s: Mrs. Thatcher's Legacy," *Journal of Management Studies* 29 (1992), 571–89, on p. 577.

24. Wyn Grant, *Government and Industry: A Comparative Analysis of the U.S., Canada, and the U.K.* (Aldershot: Edward Elgar, 1989), p. 86.

25. Keith Middlemas, *Power, Competition, and the State,* vol. 3: *The End of the Postwar Era: Britain Since 1974* (Houndsmills: Macmillan, 1991), pp. 354–55.

26. Secretary of State for Industry, *Biotechnology,* Cmnd. 8177 (March 1981), p. 3.

27. British Coordinating Committee for Biotechnology, "The Strategy for Biotechnology in Britain: Report of a Meeting Held at the Society of Chemical Industry, 14–15 Belgrave Square, London, 23 October 1981," p. 7.

28. House of Commons, "Interim Report on the Protection of the Research Base in Biotechnology: Sixth Report of the Committee on Education, Science, and Arts" (Sessional Papers, 1981–82), *Biotechnology,* 27 July 1982.

29. Trade Union Congress, *Biotechnology Policy,* position paper, 19 Oct. 1982.

30. Grant, *Government and Industry* (note 24), p. 89; and Middlemas, *Britain Since 1974* (note 25), p. 356.

31. Interview, Department of Trade and Industry, London, 8 July 1992.

32. National Economic Development Council, *Government Support for Biotechnology: A Summary of a Memorandum by the Secretary of State for Industry* (London: HMSO, 1983).

33. National Economic Development Council, *Government Support for Biotechnology* (note 32) pp. 3–4.

34. Evans, Ewing, Nolan, "Industrial Relations" (note 23), p. 584.

35. Advisory Council on Science and Technology (ACOST), *Developments in Biotechnology* (London: HMSO, 1990).

36. Thompson, *Political Economy of the New Right* (note 22), p. 68.

37. See ACOST, *Developments in Biotechnology* (note 35); EC Commission/CUBE, *National Biotech Policy: EC Member States Review,* May 1992 (Brussels), 1992; Maurice Lex, "The U.K. Science Base," in *The U.K. Biotechnology Handbook '90,* ed. Anita Crafts-Lighty, Elizabeth Burak Reed, and Shirley Lanning (Canberely: Bioindustry Association, 1990), pp. 39–42, on p. 39. All figures given are estimates based on different definitions of biotechnology and should be interpreted as rough indicators.

38. Alexander Gerschenkron, *Economic Backwardness in Historical Perspective* (Cambridge, Mass.: Harvard University Press, 1962); and John Zysman, *Governments, Markets, and Growth: Financial Systems and the Politics of Industrial Change* (Ithaca, N.Y.: Cornell University Press, 1983), pp. 171–231.

39. Andrew Cox, "The State, Finance and Industry Relationship in Comparative Perspective," in *State, Finance and Industry: A Comparative Analysis of Post-War Trends in Six Advanced Industrial Economies,* ed. Cox (Brighton: Wheatsheaf, 1986), pp. 1–59, on pp. 39–40.

40. Cox, "The State, Finance and Industry Relationship" (note 39), pp. 42–47, 35.

41. Julia Irvine, "Risk-Averse Lenders in a High-Risk Area," *Accountancy* 106 (1990), 134–35; Michael Kenward, "Little Ventured, Nothing Gained," *Director* 46 (1992), 35; and Gordon C. Murray, "A Challenging Marketplace for Venture Capital," *Long Range Planning* 25 (1992), 79–86.

42. "Stock-Exchange Proposals Disappoint U.K. Firms," *BioTechnology* 11 (Feb. 1993), 146–47; and "Does U.K. Biotech Offer Better Opportunities?" *BioTechnology* 11 (June 1993), 106.

43. National Economic Development Office, *New Life for Industry: Biotechnology, Industry and the Community in the 1990s and Beyond* (London: HMSO, 1990); and Mark Dodgson, "Strategic Alignment and Organizational Options in Biotechnology Firms," *Technology Analysis & Strategic Management* 3 (1991), 115–25.

44. Jacqueline Senker, "UK Biotechnology: Technology Transfer Involving Small and Medium-Sized Firms," *Industry & Higher Education* (June 1991), 108–13.

45. The Department of Trade and Industry (DTI) was split into the Department of Trade and the Department of Industry in 1974 and reconstituted in 1983.

46. Interview, Department of Trade and Industry, 9 July 1992.

47. Philip Gummet, "History, Development and Organisation of UK Science and

Technology up to 1982," in *Science and Technology in the United Kingdom,* ed. Robin Nicholson, Catherine M. Cunningham, and Philip Gummet (Harlow: Longham, 1991), 14–26, on p. 16.

48. Hilary Rose and Steven Rose, *Science and Society* (London: Allen Lane/Penguin, 1969), p. 24.

49. Brian Balmer and Margaret Sharp, "The Battle for Biotechnology: Scientific and Technological Paradigms and the Management of Biotechnology in Britain in the 1980s," *Research Policy* 22 (1993), 463–78, on pp. 473–75.

50. "Review of the Research Council's Responsibilities for the Biological Sciences: Report of a Committee Under the Chairmanship of J. R. S. Morris C.B.E. F.Eng., presented to the Advisory Board for the Research Councils, April 1989" (MS), pp. 24–25.

51. "British Research Councils Win and Lose," *Nature* 364 (1993), 272.

52. Senker, *UK Biotechnology* (note 44), p. 108.

53. Jacqueline M. Senker, "Conflict and Cooperation—Industrial Funding of University Research," *Journal of General Management* 15 (1990), 57–62.

Biotechnology and the Creation of a New Economic Space

Martin Kenney

The commercial potential of molecular biology and its kindred disciplines was first recognized in the mid-1970s. In the following years capitalist enterprises in the United States and abroad adopted the techniques of molecular biology, a scientific discipline. In the process, molecular biology has transformed an engineering discipline, bioprocess engineering, and spawned an industrial field, biotechnology. Biotechnology as a business arises out of an intersection of the scientific practices of molecular biology—formerly undertaken only in universities—and the engineering practices of biochemical engineering and other technologies necessary to produce biological commodities. In Joseph Schumpeter's terms, it was at this intersection that a "new space" for economic activity was created.[1]

The creation of a new economic space for biology is not unique in the history of capitalist development. There is a long history of the results of biological research being commercialized.[2] However, biotechnology presents unusual features. For the sake of brevity, this article begins with the 1970s, and I do not discuss the "mechanism" orientation of molecular biology, though some have argued the importance of this perspective to its development.[3] True, as George Basalla pointed out, any particular technology has evolved from previous developments.[4] The new biotechnology industry is built upon a base of knowledge in fermentation and biological materials processing developed in the pharmaceuticals and food processing industries.[5] Thus the commercialization phase has antecedents on several levels.

We should be extremely cautious when we term biotechnology "an industry." Only in the United States has biotechnology become an industry composed of freestanding enterprises; in other advanced industrial nations it has been subsumed under the traditional multinational pharmaceutical, chemical, and food firms and industries (hereafter, multinational companies or MNCs). For these established MNCs, biotechnology is what might be termed an "enabling" technology. Biotechnology will improve their ability to continue to operate in their current line of business as the techniques of molecular biology become central to their research and development efforts in biologically related fields. But outside the United States few new firms—and thus no new industry—have arisen. Nevertheless, for the sake of linguistic simplicity and because this article focuses on the United States, it treats biotechnology as an industry, even while recognizing the limitations of this characterization.

To operate a business in capitalism one needs a commodity that can be sold. The commodity is a crucial category, for it is what is traded in the marketplace. Not everything is a commodity, however, and things that were commodities can be removed from the realm of commodities and vice versa. For example, before the Civil War human beings of African descent were considered commodities (things) and traded.[6] Conversely, things that are not commodities can become commodities. The classic case of commodity creation is the declaration by Western colonialists that the new territories they had invaded and conquered constituted private property. The colonial authorities created a market in land, but they invariably had to use force to secure these commodity relations. Similarly, contrary to the ideology of contemporary economics, markets are not "natural"; rather, they are created by political action. These fundamental premises underlie my analysis of the creation of the biotechnology industry.

By definition, new economic spaces do not exist, but rather are created and ultimately populated by firms. To accomplish this, capital must be gathered, employees secured, legal norms and rules promulgated, and numerous other relations developed. For example, society must recognize the results of human activity in that particular area as commodities. If the results of a productive activity cannot be considered a commodity and sold in the market, private capital will not be invested. Thus industrial pioneers often must not only develop the product but create the social, legal, and economic institutions within which the product is embedded.[7]

Describing the establishment of a new economic space is difficult, because the various strands that create it are intertwined. Each strand has its own logic, which is not entirely derived from the other strands, and yet the strands also interact. For example, the development of the legal system relating to living organisms was synchronous and largely, but not entirely, related to the commercialization of biotechnology. In this sense, there is not one but rather a number of important previous strands. For this article, even though Genentech was not the first biotechnology company, since Cetus was established in 1971, Genentech's rapid success makes its establishment a convenient starting point.

This article is divided into four sections. The first examines the background in the 1970s from which the biotechnology industry would emerge. The second explores the establishment of the small start-up firms that pioneered the new biotechnology and the crucial role of venture capitalists in that process. The third section describes how burgeoning investment in biotechnology led to the creation of firms, industry associations, and trade newspapers that formed an infrastructure to support the biotechnology industry. The fourth section examines the MNC role in responses to and strategies for creating a new economic space in the form of a biotechnology industry. The conclusion reexamines the commercialization of university science from a Schumpeterian perspective. I argue that Schumpeter provides a unique perspective for examining the manner by which capitalist enterprises extend their sway into new areas of the natural world and in the process create new economic spaces.

The Background Environment

The contemporary biotechnology industry emerged from university-based research in the basic science of molecular biology. Laboratory success in the early 1970s at cloning transgenic bacteria initially sparked controversy over the safety of recombinant DNA.[8] By 1977 the safety issues had given way to arguments promising future agricultural and medical benefits of what had by then come to be called biotechnology.[9] The locus of public concern shifted to the impact that rapid commercialization of university knowledge and personnel had on the university as an institution.

Concern over the commercialization of university research has persisted because more than for any other commercialized technology, the techniques and even the target products of biotechnology were pioneered in university laboratories. Traditionally, information of interest to the pharmaceutical industries diffused from the university through the placement of students or professorial consultancies with firms. The pattern in biotechnology was quite different.

Whether living organisms and components of organisms could be patented was another important issue being settled as the biotechnology industry emerged. Whether patents were necessary for the industry's development is difficult to answer unequivocally, but the type and character of the evolving patent regime sparked great interest in its early days.[10] As Daniel J. Kevles points out in this volume, patents on biological materials are not without precedent. Still, the extension of intellectual property protection to living organisms was significant because developing and commercializing a pharmaceutical is extremely expensive. If the pharmaceutical is not patented, a rival can copy and produce an exact duplicate of the drug for far less investment than the initial inventor's. The patent allows the owner to attempt to recover costs through monopoly profits. Patents are especially important for start-up firms because knowledge and materials developed by research not only are capitalized in their stock value but often are the firm's only significant assets.

Besides deciding whether living organisms would be patentable, the federal government had to decide whether universities could patent professorial inventions. The invention that prompted the debate about the issue was Stanford University's petition to NIH in 1974 over the acceptability of its filing the basic genetic engineering patent, the Cohen-Boyer patent.[11] The NIH decision to allow universities to patent and license in the field of genetic engineering simplified the privatization of university research by removing any claims on behalf of the public regarding ownership of government-funded research.

By the early 1980s, when it became clear how much money could be made, the ethos discouraging the patenting of biological materials changed significantly.[12] Albert Halluin observes that "early discoveries [1972–77] on the construction of plasmids and vectors were published . . . but patents on such processes and compositions were not sought. [But] in recent years, patents have been sought and obtained on processes for obtaining vectors and plasmids." Halluin's conclusion can be illustrated by an example from the field of monoclonal antibody (MAb) techniques. In 1975 Cesar Milstein and Georges Köhler invented a process for creating MAbs for which they later won the Nobel Prize. They chose not to patent their invention, which became an important commercial technology. In the intervening five years the atmosphere changed drastically, and in 1980 Milstein filed for a patent on a particular MAb.[13]

With increasing commercialization, the etiquette of exchanging biological materials changed significantly. About 1984 laboratories began requiring that researchers wishing to borrow biological materials complete forms prior to fulfillment of their requests. The Harvard University biological materials supply form resembles Genentech's except for a clause on publication. Genentech's permission says "I understand I may publish the results of my experiment using plasmid —— but only with the consent of Genentech; such consent not to be untimely or unreasonably withheld." Harvard University requires only that the plasmids be used for research purposes and that the requesting party be "in periodic contact with [researcher] at Harvard . . . to report on [any] work which utilizes the plasmid." César Milstein's laboratory at the Medical Research Council had a very different form requiring only that the cell lines be acknowledged, a preprint of the paper be sent to his laboratory, and any products not be made the subject of patent rights.[14] This formalization of the conditions of exchange was a direct outcome of the privatization both of biological materials and tools and of the results of molecular biological research.[15]

Resolution of health and safety issues, recognition that the university could legitimately patent and license biotechnology inventions, and increasing acceptance of the privatization of biological materials were crucial to the creation of a biotechnology industry. And yet the most important social development was society's acceptance that publicly funded research in tax-exempt universities was appropriate for privatization. This made it possible to have a market in biological materials and know-how.

Establishing an Industry

The most basic model of capitalism is one in which an entrepreneur invests in plant and equipment and hires labor. During most of the postwar period, however, the large pharmaceutical firms with research laboratories and extensive marketing networks drastically increased barriers to entry for interested entrepreneurs. Only in the United States did an unusual set of social circumstances combine with the technological discontinuity caused by the development of new biological techniques to allow creation of a biotechnology industry based on small firms.

The technical developments of recombinant DNA technology offered the hope that cells could be transformed into "factories" for valuable biological materials and thus open up business opportunities. Curiously, the debate and publicity about health and safety issues actually attracted the attention of venture capitalists, the potential financial backers; it may also have discouraged established pharmaceutical firms from capturing the technology.

The promise and perils of biotechnology generated the most intense attention and controversy in the San Francisco Bay area and around Boston. Both regions were also centers of innovation for the burgeoning microelectronics and computer industries. As with biotechnology, the electronics and computer industries required highly trained university graduates as employees, and some firms depended directly upon university research. Thus Boston and the Bay Area became hosts to a number of extremely successful electronics start-ups that rapidly grew to be established firms and in the process created an entrepreneurial environment.[16] Most important, the economic success of both regions helped create a set of specialized financial intermediaries, the venture capitalists.

Venture capitalists are, in large measure, a set of financial intermediaries unique to the United States. Venture capitalism first arose immediately after World War II; it has since grown to become a financial sector with assets in excess of $15 billion. Organized in partnerships with ten-year limits, venture capitalists seek capital from institutions and wealthy individuals and invest it in high-risk, high-reward ventures. For the most part they have confined their lending to high-technology fields. The growth of this sector was fueled by the extraordinary high returns secured by the venture capitalists who made the initial investments in what are now Fortune 500 companies—Apple Computers, Digital Equipment Corporation, Sun Microsystems, Lotus, and Intel.[17]

Venture capitalists became involved in biotechnology at an early stage because they were already familiar with unproven technologies and willing to invest in them. Their experience with the high-technology electronics and semiconductor industries and their financial success made them comfortable with funding technologically sophisticated projects. Moreover, because the electronics industry was located near universities, the venture capitalists were also located near the university laboratories of the molecular biologists. This confluence of variables meant that biotechnology had available one of the preconditions to starting a business: a mobilizable source of capital.

Venture capital investments are quite different from traditional bank loans or equity investments. Their objective is to increase the value of the fledgling company rapidly so that the investors may sell equity to the public or to a larger corporation. To accelerate a firm's growth, a venture capitalist will help secure professional legal and accounting assistance, hire key executives, contact potential business partners, find the right underwriters for a public offering, and provide both the capital and the contacts necessary for a firm to become self-sufficient. Put another way, venture capitalists can assist an investment in transforming itself from a firm in name only to an actual operating firm. Through these activities, the venture capitalist lowers the entry barriers for entrepreneurs.[18]

The most important route for privatizing the knowledge and skills contained in the university was through the new biotechnology firms funded by venture capitalists. Venture capital financing of biology professors was first used to create a commercial firm based on the research undertaken at the University of California, San Francisco (UCSF). The company, Genentech, was founded in January 1976 by a venture capitalist who had been affiliated with the venture capital partnership Kleiner Perkins, Robert Swanson. His scientist partner was Herbert Boyer, a professor at UCSF. The business offices of Genentech were initially in the offices of Kleiner Perkins, which also made the initial $100,000 investment in the new firm. For the next two years Genentech would use Boyer's university laboratory for its experiments in cloning first a human somatostatin gene and then a human insulin gene into bacteria.[19]

Genentech's first employees were some of the postdoctoral students in Boyer's laboratory who began to work exclusively on company projects.[20] Their success was marked by Genentech's announcement in 1978 that it had cloned a human insulin gene. Eli Lilly, the largest U.S. producer of insulin, then announced that it had licensed the cloned microorganism from Genentech.[21] This transaction validated biotechnology as an endeavor that could produce a commercially interesting result. The saga of Genentech's birth was completed in 1980 when an initial public offering of its stock was made. The offering price was $35 per share, but the stock was so oversubscribed that the price per share soared to more than $80 on the day of offer, after which it fell to approximately the offering prices. This successful offering demonstrated that biotechnology companies could be successfully sold to the public even while they had negative cash flow and no products on the market. That success triggered what Nicholas Wade, writing in *Science*, called a "gold rush" of entrants.[22]

These new companies spawned rapidly, as escalating numbers of those founded reflect. From 1971 through 1978 only 19 firms were launched; 9 were established in 1979, 18 in 1980, and 33 in 1981.[23] The number of startups decreased to 11 in 1982 and 4 in 1983. Of the numerous possible explanations for this decrease, the most powerful has less to do with biotechnology than with the weakness of the market for initial public stock offerings. That is, the 1982–83 recession limited the ability of venture capitalists to sell their stock in newly established firms, forcing them to maintain their original investments in already-established biotechnology firms, which continued to lose money. The market for initial public offerings of small firms improved again in the late 1980s, and formation of new firms once again accelerated. According to a recent estimate there are now nearly 700 biotechnology firms in the United States.[24]

In the early phase of the industry there was no pool of expertise in industrial biotechnology per se. Both the managers and the technical employees to staff the start-up companies had to be drawn from other sectors of the economy. The obvious source for technical staff was the university, where a large pool of postdoctoral students could be recruited. More difficult was persuading university scientists to leave tenured, well-paying positions for the private sector; persuading the top scientists was often impossible. To circumvent this problem, the start-ups developed a unique structural feature, the scientific advisory board, touted as a "scientific board of directors" and usually consisting of prominent scientists from major universities. The company literature described their role as advisers, recruiters of trained personnel, and information sources on current developments in academic science.[25] In return for participating in the scientific advisory board, these scientists received significant sums of stock. Later, several became multimillionaires when the company went public.

The new firms also needed managers familiar with the pharmaceutical industry; they secured them through such normal channels as advertising and executive search. The start-ups had little difficulty securing these trained personnel because they could offer stock options and rapid promotion. Thus the start-ups were able to create both managerial and scientific teams quite rapidly and begin operations. By the 1980s it was even possible to secure personnel with experience in the biotechnology industry by raiding other companies and increasing numbers of graduates sought employment in the now established industry. The new economic space had developed its own labor market, distinct from the university or the pharmaceutical industry.

Creating an Infrastructure

A new technology or industry does not and cannot exist in a vacuum. Clearly, a scientific enterprise also purchases inputs such as laboratory ware, scientific equipment, and other consumables. The creation of NBFs and the acceleration of spending in corporate research budgets fueled the development of an infrastructure of biotechnology input firms. In other words, backward linkages were rapidly built. Thus, for example, companies such as Applied Biosystems were established to produce machinery, reagents, and other inputs to the biotechnology industry. The new infrastructure then reinforced the capabilities of the biotechnology firms and speeded procedures such as sequencing and cloning genes.

Another indicator of the growth of the biotechnology industry was the rise of trade journals that knit the industry together and provided it with a voice. Before biotechnology was commercialized, peer-reviewed scholarly journals such as *Science* and *Nature* constituted the communication medium for molecular biologists. As biotechnology grew in the late 1970s, demand for business information increased, and a number of newsletters appeared. In 1981 the first self-conscious "industry" trade newspaper, *Genetic Engineering News* (*GEN*), was published. *GEN* has a subscription fee but is largely supported by advertising revenue from biotechnology input suppliers. Biotechnology's growth as an industry can be traced in *GEN*'s publication schedule: In 1981 it was bimonthly, by 1987 it was monthly, and in 1992 it became biweekly. Changes in its subtitle reflect the growing complexity of the industry. In 1981 it was "The Information Source of the Biotechnology Industry"; in 1987 it had changed to "The Source of Bioprocess/Biotechnology News"; and in 1992 it changed yet again to "Biotechnology, Bioregulation, Bioprocess, and Bioresearch." *GEN* was followed in 1984 by the British journal *Bio/Technology* and a French journal, *Biofutur*. The latter two journals depended more on subscriptions, but also had a significant number of advertisements by input manufacturers. Significantly, scientific labor continued to be recruited through the classified advertisements in *Science*. In effect, *GEN* was supported by the biotechnology input industry and became a voice for commercial biotechnology.

The growth of an industry can also be traced through the development of its industry associations. During the earliest days the Pharmaceutical Manufacturers Association was the de facto voice of the industry. In 1981 seven new firms combined to charter the Industrial Biotechnology Association (IBA), and membership quickly grew to eleven firms.[26] In 1984 Mary Ann Liebert, Inc., the publisher of *GEN*, and eleven other companies joined to form the Association of Biotechnology Companies (ABC), with the express purpose of representing the smaller biotechnology firms, some of which did not believe the IBA was articulating their needs.[27] In 1993 these two organizations, the IBA with 150 members and the ABC with 340 members, merged to form the Biotechnology Industry Organization. The new organization would represent firms with $5.9 billion in sales and the majority of the estimated 79,000 jobs in the biotechnology industry.[28]

The MNCs and the New Biotechnology

The established chemical and pharmaceutical MNCs constitute the other set of firms participating in the privatization of molecular biology. These firms have multibillion dollar revenues and are members of an established global industry. In the early 1970s, as the recombinant DNA controversy was under way, the Pharmaceutical Manufacturers Association was the most significant industrial lobbying group. At the time, however, its constituent pharmaceutical firms did not have significant internal expertise in molecular biology and especially in recombinant DNA. The MNCs initially thought of these techniques as research tools and not as potential generators of products. In general these companies were not able to integrate biotechnology quickly into their research programs. Thus a technological discontinuity emerged that, at least momentarily, lowered the entry barriers into the pharmaceutical industry.

Genentech's success at cloning a human insulin gene and Lilly's 1978 purchase of the rights to the microorganism created the possibility that the new biotechnology could outflank existing products and possibly lead to lucrative new products. The MNCs responded by following three strategies. First, they established their own internal biotechnology research programs in molecular biology. Second, they established their own linkages to university laboratories. Third, they developed strategic partnerships with the small start-up firms. Each of these strategies had advantages.

The MNCs' internal research programs had a number of significant early problems. Most important was difficulty in recruiting the best scientists. The reasons for this varied, but the one most mentioned is that the companies were not undertaking cutting-edge research. Moreover, these companies did not have the networks necessary to recruit the top molecular biology graduates, nor were their laboratories located near the top universities, as the start-up firms were, allowing easy recruitment. Even today the MNCs are still not considered to have the topnotch biotechnology researchers in their laboratories.

In the early 1980s the MNCs funded a number of long-term agreements with universities aimed at gaining access and intellectual property from academic laboratories. Perhaps the most publicized was the agreement made between the German chemical-pharmaceutical giant Hoechst and the Massachusetts General Hospital. By giving the hospital roughly $70 million, Hoechst was able to purchase access to the intellectual property of a newly constituted department. In the late 1980s these large, long-term arrangements tapered off, though occasional examples still occur, such as the agreement signed between the Hitachi Chemical Corporation and the University of California at Irvine in 1990. Undoubtedly, smaller arrangements continue to be signed between MNCs and university laboratories, but the trend appears downward. University patenting and licensing of the inventions of professors, on the other hand, continues to increase.

The dominant trend is for the small biotechnology firms to conclude cooperative agreements with the MNCs, because the skills and resources it takes to develop a new drug or diagnostic are quite different from those necessary to take a drug through clinical trials and FDA approval and to market it to doctors. The smaller start-ups can develop the new product, but the costs and personnel needed to secure FDA approval are often beyond their resources. The U.S. biotechnology industry consisting of small firms, thus depends for funding and certain types of skills on the MNCs. The difficulties of becoming a truly independent freestanding pharmaceutical firm are plain. Genentech, for example, one of the two most successful start-ups, sold 60 percent of its stock to Hoffmann-La Roche. Hybritech, one of the more successful MAb companies, was sold to Eli Lilly in 1985. There are very few start-ups that do not have at least one contract with an MNC.

The MNCs have successfully used their financial resources, skills in dealing with the regulatory process, and superior marketing expertise to secure access to developments in the new biotechnology. However, the MNCs have not been able to internalize the knowledge necessary to dispense with either university research or, especially, the start-ups. The MNCs therefore continue to contract with the start-ups for novel products or processes, and this keeps the new economic space separate from the pharmaceutical industry.

Conclusion

The context of biological research has changed markedly in the last two decades. In the late 1970s and early 1980s the commercialization of biological knowledge centered on the technologies of recombinant DNA and monoclonal antibodies. By 1993 firms had been formed to exploit liposomes, antisense molecules, peptides, carbohydrates, stem cells, synthetic small molecules, and many other biological materials. The new economic space opened by the small firms in the early and mid-1970s became an important field for profit making and continues to grow. The persistent predictions of a shakeout and reconsolidation into the pharmaceutical industry are based on the assumption that the new economic space will collapse. Yet some of the established biotechnology firms continue to survive and to progress toward commercializable products. Moreover, there continues to be a constant (though cyclical) flow of NBFs. Even as some firms are absorbed, new ones are formed.

Clearly, the biological sciences have undergone privatization in the last twenty years. But what is striking is that the public science continues to flourish and generate new possibilities for private appropriation. In this sense, the economic space continues to expand. Conversely, the partial reassertion of the MNCs indicates that though a new economic space has been created, it has not become entirely independent. Perhaps this can be understood by returning to our earlier bracketing of the debate as to whether biotechnology is an industry or an enabling technology. If biotechnology is an industry, then it should have developed a discrete economic space. If it is merely an enabling technology, then it should not be able to build space.

An organized venture capital community was crucial in opening this new economic space. In Europe and Japan, which did not have a venture capital community, biotechnology was reduced to practice in the large established chemical and pharmaceutical firms.[29] The institutional context had an important influence on the characteristics of the newly created economic space. Analysis of the socioeconomic institutions in which molecular biology and later biotechnology were and are embedded provides an important case study of how scientific developments in the public domain can be translated into private sector commodities.

There can be little doubt that biotechnology is now, in some measure, a private science. This constrained the free flow in biological materials and information not only among private sector entities, but also in the public sector. The "public goods" context upon which free flow was predicated, which existed prior to the commercialization, has changed fundamentally: what was formerly science is now technology. Similarly, the motives and conditions of all of the social actors in the field of biotechnology have changed to adjust to this privatization.

Notes

1. Joseph Schumpeter, *Business Cycles: A Theoretical, Historical and Statistical Analysis of the Capitalist Process*, abridged by R. Fels (New York: McGraw-Hill, 1939).

2. See, e.g., Jack Kloppenburg, *First the Seed: The Political Economy of Plant Biotechnology* (New York: Cambridge University Press, 1987); Bruno Latour, *The Pasteurization of France* (Cambridge, Mass.: Harvard University Press, 1988); and the article by Angela Creager in this volume.

3. See Edward Yoxen, "Life as a Productive Force: Capitalizing the Science and Technology of Molecular Biology," in *Science, Technology and the Labour Process: Marxist Studies*, ed. Les Levidow and Bob Young (Atlantic Highlands, N.J.: Humanities Press, 1981), pp. 66–123; and Pnina Abir-Am, "The Discourse of Physical Power and Biological Knowledge in the 1930s: A Reappraisal of the Rockefeller Foundation's 'Policy' in Molecular Biology," *Social Studies of Science* 12 (1982), 341–82.

4. George Basalla, *The Evolution of Technology* (Cambridge: Cambridge University Press, 1989).

5. R. K. Finn, "Some Origins of Biotechnology," *Swiss Review for Biotechnology* 7 (1989), 15–17. See also the article by Robert Bud in this volume.

6. The word *things* is used very broadly here. For the fundamental discussions of

what a commodity is, see Karl Marx, *Capital*, vol. 1. For further insightful theoretical discussions of the nature of commodities, see Kloppenburg, *First the Seed* (note 2).

7. Although I agree that the marketplace is clearly created, I do not take the radical subjectivist position that the commodities developed by biotechnology are the subjective creation of society, or that their efficacy is entirely an intersubjective creation. That position if reduced to the absurd would intimate that anything agreed upon socially would be "effective."

8. For a comprehensive discussion of the health and safety issues underlying the recombinant DNA controversy, see Sheldon Krimsky, *Genetic Alchemy: The Social History of the Recombinant DNA Controversy* (Cambridge, Mass: MIT Press, 1982). See also David Jackson and Stephen Stich, eds., *The Recombinant DNA Debate* (Englewood Cliffs, N.J.: Prentice-Hall, 1979). For a readable popular account see John Lear, *Recombinant DNA: The Untold Story* (New York: Crown Publishers, 1978). For further discussion see the article by Susan Wright in this volume.

9. Krimsky, *Genetic Alchemy* (note 8); and Martin Kenney, *Biotechnology: The University-Industrial Complex* (New Haven, Conn.: Yale University Press, 1986). See also the article by Susan Wright in this volume.

10. For a more general argument about the importance of patents see Eric Schiff, *Industrialization Without Patents* (Princeton, N.J.: Princeton University Press, 1971).

11. The Cohen-Boyer patent covers the insertion of foreign genes into a plasmid and then the transference of those genes to another microorganism.

12. Jorge Goldstein, "A Footnote on the Cohen-Boyer Patent and Other Musings," *Recombinant DNA Technical Bulletin* 5 (1982), 180–88, on p. 186.

13. Albert Halluin, "Patenting the Results of Genetic Engineering Research: An Overview," in *Patenting of Life Forms*, ed. David W. Plant, Niels J. Reimers and Norton D. Zinder, Banbury Report 10 (Cold Spring Harbor, N.Y.: Cold Spring Harbor Laboratory, 1982), pp. 67–126, on pp. 75, 79ff.

14. Compare Genentech, "Letter Agreement/Biological or Other Materials Supply" (ca. 1984); Harvard University, Committee on Patents and Copyrights, "Plasmid Supply Letter" (ca. 1984); and César Milstein, "Cell Line Supply Form" (Cambridge, Mass.: MRC Laboratory of Molecular Biology, ca. 1984).

15. For a provocative analysis of the development of patenting in biological materials, see the article by Alberto Cambrosio and Peter Keating in this volume.

16. There are important differences as well as similarities between the biotechnology and electronics-based industries. Industry laboratories accomplished many of the central developments in computers and microelectronics, and professors were much less involved in establishing and managing electronics industry firms. Overdrawing the parallels can lead to misunderstanding the development of the two industries.

17. Richard L. Florida and Martin Kenney, *The Breakthrough Illusion: Corporate America's Failure to Move from Innovation to Mass Production* (New York: Basic Books, 1990). See also John Wilson, *The New Venturers: Inside the High Stakes World of Venture Capital* (Reading, Mass.: Addison-Wesley, 1985).

18. For further discussion see Florida and Kenney, *Breakthrough Illusion* (note 17).

19. See Lear, *Recombinant DNA* (note 8), pp. 232–34; and Laurel Glass, chairman-elect, Academic Senate, memorandum to Keith Yamamoto, University of California, San Francisco, 30 Aug. 1979. I thank Keith Yamamoto for providing a copy of this memorandum.

20. Kenney, *Biotechnology* (note 9).

21. Krimsky, *Genetic Alchemy* (note 8), p. 201.

22. Nicholas Wade, "Cloning Gold Rush Turns Basic Biology into Big Business," *Science* 208 (1980), 688–92.

23. Kenney, *Biotechnology* (note 9).

24. "Why Are We So Afraid of Growth?" *Business Week*, 16 May 1994, pp. 62–72.
25. Kenney, *Biotechnology* (note 9).
26. P. Germann, "IBA Gets Underway," *Genetic Engineering News*, Nov.–Dec. 1981, p. 6.
27. "Small Biotech Firms Form a Trade Group," *Chemical Week*, 25 Jan. 1984, pp. 13–14.
28. E. Christensen, "The Biotechnology Industry Organization: The Sum Is Greater than the Parts," *Genetic Engineering News*, 1 Apr. 1993, p. 17.
29. For discussion of Europe, and especially the United Kingdom, see the article by Herbert Gottweis in this volume. On Japan see Malcolm V. Brock, *Biotechnology in Japan* (London: Routledge, 1989).

The Cultural and Symbolic Dimensions of Agricultural Biotechnology

Sheldon Krimsky

The application of biotechnology to agriculture is beginning to affect the deep symbolism we associate with contemporary agrarian culture. In the political discourse emanating from the "public wars over biotechnology," several key ideas are being debated: the notions of what the farm is and what it ought to be, what food is, and what livestock are.[1] Combatants in these struggles include small-farm advocacy organizations, start-up companies, international Green organizations, traditional environmental groups, multinational corporations, natural food associations, and animal rights supporters. The issues are also diverse: a tomato with an inverted gene to slow the ripening process, a microorganism that inhibits frost formation on crops, herbicide-resistant crops, microbially produced animal growth hormones, animals that excrete pharmaceuticals in milk, and transgenic animals that yield leaner or breastier meat. The common thread linking these product controversies is a set of genetic technologies that offer highly characterized genetic modifications of bacteria, plants, and animals.

The controversies over individual biotechnological products represent what might be called the *surficial* phenomena of a deeper set of issues that reach beyond the biotechnology industry and tap into a cultural immune system that responds to new product technologies. On a superficial level the global debates over biotechnological products are about controlling the system of regulation, the conditions of technology transfer, and the criteria used to approve technology for commercial application. At the deeper level, the struggle occurs among forces that seek to control the symbolism of agricultural biotechnology and seek to promote a techno-mythology based almost exclusively on the power of genes. Those who eventually gain control over the bio-mythology will affect the pathways of innovation for future generations.

This essay is about the struggle for control over the symbolic meaning of biotechnology. It is also about the process of mythmaking for this class of technologies. I use the term *myth* judiciously, not to denigrate one side or another in the battles that have erupted over biotechnology. Rather, the term signifies a cultural story that embodies hope, expectations, moral attitudes, fears, or positive visions of modernity. Myths are constructions out of reality that transcend the real into a virtual world of expectations. I will examine the following myths and anti-myths in their socioscientific context:

- Biotechnology gives us natural (unnatural) products.
- Biotechnology offers us more (less) control over nature.
- Biotechnology will contribute to more (less) biodiversity.
- Biotechnology will be friendly (unfriendly) to the environment.

These propositions are not necessarily independent of one another (e.g., being friendly to the environment should be consistent with increased biodiversity) but are useful as organizing schemas for the analysis of debates. I examine the first three propositions seriatim in the next three sections, discussing the fourth in the context of the first three. I then discuss the shifting conceptual boundaries of agricultural and industrial production and conclude by examining the form and function of "power myths" in relation to genetic technology. My organizing thesis is that political debates in biotechnology are essentially about control over techno-mythmaking, which I define as the shaping of social expectations through the association with technology of symbolic powers and simple moral virtues. I am less concerned about the truth or falsity of these techno-myths than I am about how they are used to construct an image of biotechnology.[2]

Biotechnology and the State of Nature

Former U.S. President Ronald Reagan once commented that trees emit pollutants.[3] As an anti-regulator, Reagan meant not to imply that trees must be controlled, but rather to call into question the distinction between the natural and the unnatural, and the uncritical association between the natural and the good. Despite Reagan's ploy, most people today feel comfortable with the natural-unnatural dichotomy and hold a deeper skepticism about the unnatural. It is not surprising, therefore, that representatives of the biotechnology industry have described the techniques associated with genetic modification and the resulting pesticides as offering a natural alternative to agrichemical products. By advancing the thesis that biotechnology uses nature's own methods of pest control, plant fertilization, or toxic waste degradation, its advocates hope to give biotechnology a more favorable image while disassociating biological agents from the negative image often ascribed to chemicals.

Not all representatives of the industrial community (e.g., large chemical companies) find it in their interest to pit biological agents against synthetic chemical agents. But the small venture capital companies advance this theme in part to win public confidence in genetic technologies and also to create receptive market alternatives to the chemical inputs produced by a few leading corporations.

Efforts to gain control over the term *natural* can be observed in other policy debates. Public anxiety over synthetic additives to food has been met with studies indicating that "natural carcinogens" in food are more hazardous to the consumer than traces of food additives and pesticides.[4] If this argument were taken to its logical conclusion, food products would not be regulated differently whether ingredients were indigenous to the plant, taken up in the roots, sprayed on, or added in processing. All products, whether naturally found or synthetically produced, would be on an equal footing. Only demonstrated risk, not methods or generic materials of production, would be subject to regulation.

The U.S. Food and Drug Administration (FDA) also grappled with the natural-unnatural debate when it issued draft guidelines for regulating transgenic food which stated that inserting a foreign gene into a crop did not make the crop unnatural. The agency decided not to regulate genetically engineered food in the same way as it would food to which synthetic chemicals had been added. Under the FDA's proposal a transgenic plant is considered no less natural than a plant that has been selectively bred: the difference is that a bred plant has many undefined genes making up its phenotype while a transgenic plant usually has a well-characterized genetic modification.[5]

The critical discourse of biotechnology encompasses efforts to stigmatize genetic engineering as unnatural as well as efforts to classify it as a natural process. Critics of the FDA approach, who maintain that transgenic plants are not simple extensions of selective breeding, emphasize that genetic engineering makes it possible to exchange foreign genes (a gene from a peanut transferred to a tomato) that would otherwise not be introduced by selective breeding or hybridization. But what makes hybridization natural? Could a hybrid seed arise from natural processes not subject to human intervention? The U.S.-based Foundation on Economic Trends has organized a Pure Food Campaign that opposes all genetic manipulation of plants. At least one argument of the campaign is based on the unnaturalness of transgenic food.

Another struggle for control over the symbol "natural" can be found in the controversy over agricultural pesticides. In agriculture the term *natural*, which is synonymous with "organic," has been identified by the term *chemically free*. Despite efforts by certain sectors in society to break the link between *natural* and *safe*, these concepts are still strongly associated. Food produced on chemically free farms may contain fungi, insect viruses, and natural carcinogens. Nonetheless, in the public mind organically grown food has a strikingly powerful grip over the terms *natural* and *safe*.

Among the boldest promises of biotechnology—and one that is repeatedly emphasized in the popular scientific press—is that its products will replace synthetic chemical pesticides and fertilizers with biological controls.[6] For the antichemical pesticide lobby, fulfilling this promise would place biotechnology on the side of the angels. Evidence for these claims comes directly from laboratory experiments on pest resistance, nitrogen fixation, and viral resistance in plants. The mythmaking behind this view is that enough power exists in biotechnology to displace a system of chemically intensive agriculture that has evolved for over a hundred years. The idea that biotechnology will wean agriculture from its chemical dependence is widely accepted and cautiously promoted despite four major constraints on the realization of this goal: technology, economics, education, and politics.[7] The products replacing chemical pest controls will have to match or exceed existing efficacy criteria. In addition, the economic benefits must equal or outweigh the costs of using chemical controls. Farmers need to know how to use alternative products and practices effectively. And finally, a powerful chemical lobby must be overcome as biological methods and a softer path to pest control are introduced.

While the FDA has sought to treat genetically modified food as it would conventionally bred crops, the U.S. Environmental Protection Agency (EPA) has a policy that distinguishes transgenic organisms from natural substances. Under current EPA practices, transgenic organisms other than pesticides would be regulated under the Toxic Substances Control Act (TOSCA), a major piece of gap-filling legislation passed in 1976. TOSCA's requirement of pre-manufacture notices for all new chemicals has been interpreted to include genetically modified microorganisms. The EPA commitment to regulating transgenic organisms as chemical substances further blurs the distinction that some proponents of biotechnology have made between nature's "biological friends" and its "chemical enemies."

The importance of the symbol of nature's biological friends is exemplified in the bioremediation industry. Conventional microorganisms have been used successfully in sewage treatment plants and with less success in degrading toxic chemicals in situ. The industry is composed mainly of companies that use natural strains of microorganisms screened for their efficiency to degrade a selected substrate. Few companies in the U.S. bioremediation industry have shown an interest in developing genetically engineered strains. Many are concerned that the biotechnological methods may tarnish the image of an industry that exploits the earth's natural recyclers, namely, microorganisms that break down chemicals into simpler elemental forms. Unless the biotechnology industry can remove the stigma of genetically engineered microorganisms as a product of a new and untested technology, the field of bioremediation will move with extreme caution into genetically engineered strains.

Industrialists are also seeking to exploit the so-called natural qualities of microorganisms in the area of food additives. The public is strongly apprehensive of artificial chemical additives. Susan Harlander notes: "Consumer concern regarding chemical additives in foods and consumer demand for 'natural' products have resulted in demand for microbial metabolites which can be used as natural ingredients in foods."[8] Currently, microbially produced additives such as xanthan gum and monosodium glutamate come from natural isolates. If the food industry seeks to expand the manufacture of microbially produced additives, it will have to persuade the public that genetically engineered "food-grade" microorganisms will also yield natural ingredients to ensure that these products are received more favorably than synthetic chemical additives. The pharmaceutical industry has used an analogous strategy when advertising microbially produced human insulin as a natural alternative to porcine or bovine insulin.

The debate over biotechnology as natural or unnatural will not be resolved by empirical study any more than will the debate over hybrid corn. The symbolic importance of *natural,* however, has more to do with risk than with the word's essential meaning. Once the issue of risk is managed, the question of naturalness will slowly disappear. Considerable effort has gone into communicating to the public the idea that bioengineered products inherently have no greater risk than non-bioengineered products. A report of the National Academy of Sciences (NAS) in the United States addressed the issue of unique hazards of recombinant DNA research but finessed the question whether scientists could create something totally unique to nature. An inferred connection certainly exists between unique risks and the unnatural. If we had a technology that was safe by universal criteria (e.g., passive solar energy), the issue of whether it was natural or unnatural is not likely to be of concern. In the struggle over symbols the NAS report represented a major victory for the biotechnology sector. Although the report was somewhat tentative, it contained enough language to serve the interests of a war-torn industry that had often been forced into a defensive posture to justify its existence. "There is no evidence that unique hazards exist either in the use of R-DNA techniques or in the movement of genes between unrelated organisms."[9] I have argued elsewhere that the NAS conclusion about unique hazards has less to do with good science than it does with political correctness within the scientific fraternity.[10]

Greater Control over Nature

One traditional measure of human progress is the degree to which we can control, accommodate to, or survive the forces of nature. Humans have overcome the gravitational force field of the earth, exercised some control over swelling rivers, redefined the landscape by making artificial lakes, rerouted natural water flows, drained wetlands, and irrigated deserts. Yet a growing number of people argue that progress is not made by subduing or dominating nature. They reject the premise that humanity's goal is to make nature more rational—that is, more predictable and more controllable.

Many others believe that the new science of molecular genetics offers qualitatively greater degrees of human control over natural processes while releasing a new bounty of benefits to civilization. How much can we extract from nature without paying a price? Does genetic engineering give us more control, or is that an illusion? And if biotechnology does offer more control, is it desirable to exercise that control?

In agriculture the targets for increased control are weeds, insects, temperature fluctuations, and instability of water supply. For biotechnology to offer greater control over agricultural production, it must help to reduce the erratic effects of these external impediments while sustaining high productivity. Herbicide-resistant crops (HRCs) may illustrate this point. In theory at least, HRCs can improve weed control by enabling farmers to exercise postemergent herbicide treatment that does not harm the crop. HRCs also allow farmers to rotate crops with different herbicide tolerances. Proponents of HRCs maintain that they offer a more predictable and dependable regime of weed control that is friendlier to the environment since they allow industry to shift from higher-toxicity herbicides (e.g., triazines) to lower-toxicity herbicides (e.g., glyphosates and sulfonylureas).

The battle lines over HRCs were taken up early, while companies were still seeking permits to field-test transgenic plants. Leading the cause against HRCs in the United States was the Biotechnology Working Group, a coalition of influential environmental public interest organizations. Their publication of a widely circulated report titled *Biotechnology's Bitter Harvest* set the parameters of the environmental offensive. The report excoriated the biotechnology-chemical industry for initiating a vigorous research and development effort to create new herbicide-resistant plant species. In effect, the report argued, herbicide-tolerant crops would lead farmers to depend more on all types of herbicides, help spread herbicide-resistant weeds, increase ground and surface water contamination, expose applicators and farm workers to more toxic chemicals, increase herbicide residues in food, and contaminate ecosystems. Overall, the report condemned the industry initiative, accusing its proponents of a cynical disdain for the goals of a sustainable agriculture.

> It is inescapable that the widespread use of herbicide-tolerant crops and trees will prolong the use of chemical herbicides for weed control. . . . From a social and economic standpoint, the introduction of herbicide-tolerant crops could exacerbate trends toward economic concentration in agriculture, the decrease in farm numbers, and the deterioration of rural communities. Applied in the Third World, such

plants could have unwelcome impacts on human and environmental health and genetic diversity, as well as increasing petrochemical dependence.[11]

Predictably, the reaction from industry was combative. Significant efforts were being made to craft an image of biotechnology as nature's alternative to chemical pesticides. Corporate CEOs spoke optimistically of the new synthesis of environmentalism and agriculture.[12] The critics of HRCs exploited the difficulty of fitting these products into the newly constructed image, thus undermining industry efforts to control the symbols of biotechnology as environmentally and economically friendly. In response, supporters of HRCs dismissed the arguments of environmentalists as "unscientific" and ideological. The high-profile environmental magazines carried the message that the biotechnology industry had hoped to avoid, namely that companies were engineering herbicide resistance into all major crops with the upshot of more chemical dependence.[13]

The subtle details of the controversy over the environmental impact of HRCs are linked to the broader struggle over the importance of herbicide resistance in the future of agriculture. Two distinct images of the farm portray herbicide-resistant crops as having either a rich or a dim future in food production. Rooted in the texts of Judeo-Christianity and later reinforced in the post-Baconian scientific Enlightenment, the dominant view of nature is that humans must control the wild and irrational forces of their environment. Not only is such control obligatory if we are to exploit nature's largess at highest efficiency, but there are aesthetic, moral, and religious justifications for exacting it. In brief, when nature is rationalized, it looks more pleasing, it better serves social interests, and according to Western religious doctrine it rewards the Creator who has implored humans to harness nature's secrets and to subdue its irrational impulses.

An alternative vision of the farm is that of an organic and dynamic ecosystem that cannot function under the type of mechanistic control found in industrial manufacture. In the agricultural system humans must still understand their role as one among other species living in balance. Pests are an inevitable part of the farm landscape. Farmers must learn to cohabit the land while reaping a generous but not overly selfish yield. Efforts to eradicate unwanted intruders totally in the growing area have generally failed. Once the mindset is changed from eradication to management, a more ecocentric approach to agriculture is possible. John H. Perkins distinguishes between the Total Population Management and the Integrated Pest Management paradigms, both developed in the 1950s.[14] The former held that annihilation of certain pests was within technological capability; the latter was content to maintain pest populations below certain economic thresholds. These two visions of the farm portray vastly different roles for HRCs.

Among the current justifications for greater weed control are fears over food security. Since the world population increases by 90 to 100 million people per year, Malthusian logic tells us that if we do not increase agricultural yields, there will be greater political instability due to scarcity of basic food.[15] Les Levidow argues that industry has embraced food security as the raison d'être for "total systemic control which chemical-intensive methods failed to achieve."[16] The failure of chemical control methods has been documented by David Pimentel and his coauthors: "Despite the widespread use of pesticides in the United States, pests (principally insects, plant pathogens, and weeds) destroy 37% of all potential food and fiber crops. . . . Although pesticides are generally profitable, their use does not always decrease crop losses. For example, even with the tenfold increase in insecticide use in the United States from 1945 to 1989, total crop losses from insect damage have nearly doubled from 7% to 13%."[17]

And why should biotechnology offer the prospect of eradicating pests or improving the efficiency of pest control that has eluded the chemical industry? The distinction made between internal and external controls has instilled in humans a new arrogance that nature can be brought under our control. To control nature externally is to apply physical force, chemical modifications, or energy supplements on the existing agricultural system. In contrast, internal controls involve modifying the genetics of plants to establish a more efficient system of food production with greater accommodation to external variations; for example, producing virus- or insect-resistant plants by introducing genes that immunize the plant or emit a protein toxic to the insect. Since HRCs are used in conjunction with synthetic chemicals, they exemplify applying internal and external controls simultaneously.[18]

The prospect of controlling agricultural yield through the genome of plants evokes among scientists a greater power over nature than that of manipulating external factors. Part of the reason for this attitude is the success of scientific reductionism. Explaining a phenomenon by reducing it to its smallest components remains the ultimate objective of contemporary science. Elementary particle physics, which for all practical purposes has no commercial utility, is among the most revered of the scientific enterprises, garnering billions of dollars to explore the smallest particles of nature. It is perhaps part of the general scientific ethos inherited from the philosophy of physics that prediction and control are linked, and that the power of prediction is heightened as one moves to smaller and smaller units of analysis.

Does biotechnology offer the possibility of greater control over insect resistance than is achievable with chemical pesticides? That insect populations are highly adaptable to chemical pesticides is well recognized. George Georghiou notes that at least 504 insect species have developed resistance to one or more chemicals.[19] On the basis of what is known about insect ecology, Fred Gould concludes that "there is no reason to expect that insects will not be able to adapt to these biopesticides."[20]

Ironically (as Harlander notes) a radically different system of agriculture may offer the solution to the pesticide dilemma, for example, massive tissue culture factories where there are no pests and therefore no need for chemical pesticides. "Plant cell culture used for the production of natural food ingredients offers several distinct advantages over extraction of these components from plants. Seasonal variations, unfavorable weather conditions, and epidemic diseases are not problems when plant tissue is grown under well-defined and controlling laboratory conditions.[21]

Critics of the hypothesis that biotechnology will offer greater control over nature point out that the concept of control is deceptive. One may succeed at controlling a single variable while losing control over other variables. Nevertheless, an appealing if not fanciful notion is that by manipulating a few choice genes, we will gain greater dominance over complex interactions among plants, insects, and environmental factors. (This idea is discussed further below, the section on "Genetic Power and Techno-Mythmaking.")

Biotechnology and the Diversity of Nature

The United Nations Conference on Environment and Development, held in Brazil in 1992, brought considerable attention to the question of biological diversity. The Convention on Biological Diversity was adopted by ninety-eight countries. Noticeably absent as a signatory nation was the United States. The Bush administration opposed the convention on the grounds that it would undermine U.S. patent protection for the biotechnology industry. A specifically targeted concern of the treaty was that "the original possessor of naturally occurring materials can assert an interest in derivative materials after allowing them to leave their possession."[22] Although the controversy was over intellectual property rights and the ownership of genetic resources, its outcome was that biotechnology and biodiversity were opposed.

Concerns about the connection between patent rights and the loss of biodiversity were first raised in the mid-1980s when it was noted that small seed companies were being bought out by multinational corporations like Monsanto and Ciba-Geigy that were engaged in developing both seeds and agrichemicals.[23] Opponents argued that patenting genetic resources results in fewer plant varieties and in the control over those varieties by a group of oligopolistic enterprises. Michael Fox is among those voicing skepticism that biodiversity is compatible with large-scale industrial interests in biotechnology: "There is an accelerating loss of biodiversity caused by agribusiness's overreliance on a few utility strains and varieties of seed stock and livestock."[24] And in Henk Hobbelink's view the outcome of patent rights for germ plasm would mean "the total loss of genetic diversity that is maintained in the field by farmers through the selection and use of their own seed."[25]

Ironically, representatives of the biotechnology industry have actively promoted the new industry as inherently compatible with biological diversity, particularly since the symbolism of biodiversity has gained a prominent place in public consciousness. Environmentalists have invoked biodiversity as a standard to which agricultural biotechnology must measure up. The battle over control of the symbolism has been particularly complex because so many different aspects of biodiversity are involved. The message emanating from the community of international development nongovernmental organizations (NGOs) is that biodiversity is a political and not a technical problem and therefore could never be solved by biotechnology. The practice of biological diversity can only be ensured through decentralization. Centralized systems of research, production, or conservation force the spread of genetic uniformity and genetic erosion.[26]

Another concern linking threats to biodiversity with the release of genetically engineered plants is that the latter could run amok in the environment, overtaking other plant species, particularly those already on the endangered species list. Faced with a potential displacement of species, worldwide attention has focused on the risks of releasing transgenic plants into the environment.[27]

Industrial interests have vigorously promoted biotechnology as compatible with and nurturing of biodiversity. Some of their arguments have persuaded mainstream environmentalists.[28] One argument is that improving agricultural productivity through biotechnology will enable more food to be grown on less land, releasing more land for other purposes.[29] This all-purpose argument could apply to any technology that improved productivity, even chemical technology or innovations in high-yield seeds. Increases in the world population could wipe out any biodiversity benefits derived from higher productivity. The argument also assumes that by improving productivity and narrowing diversity in agricultural land we increase or maintain biological diversity in the fields left fallow.[30] The argument would be more credible if there were evidence that the improvements in production per hectare resulting from technological innovations (e.g., high-yield strains, mechanization) increased biodiversity in land protected from cultivation.

One final link between biotechnology and biodiversity is uniquely associated with the creation of novel species. New techniques that enable scientists to create intergeneric crosses of plants, animals, and microorganisms, promise an artificially created biodiversity. This initiative may take several forms. Substitution of hardier strains of plants and animals able to survive new environments affected by human development represents one path. Also, plant and animal germ plasm can be banked for withdrawal if a crisis of extinction arises. The fictional dinosaur park in *Jurassic Park* offers an extreme view of an artificially constructed ecological system reinstituting extinct species. The symbol of Jurassic Park is precisely what the biotechnology industry seeks to avoid: the introduction of uncontrollable and possibly irreversible risks in an effort to create greater diversity.

Agricultural-Industrial Inversion

The rapid industrialization and corporatization of farming in the last fifty years does little for the image of the farm as the embodiment of pastoral life. Despite the transformation of agriculture and the decline of agrarian culture in many industrial nations, the farm is still the place where germ plasm, sun, water, and earth are transformed to basic foodstuffs. With the exception of certain fruits and vegetables, most food undergoes some processing before it reaches the marketplace.[31] The trend is toward fewer unprocessed crops. Even apples are routinely sprayed with pesticides and chemical ripening agents and cucumbers are waxed before going to market.

The distinction of farm and city has always been a central theme in American cultural history. While we understand that modern large-scale farms are not pristine habitats and well-balanced sustainable ecosystems, there is still enough in the practice of farming, particularly its dependence on sunlight, soil, plants, and animals, to distinguish it from industrial manufacture, characterized by a closed system of production and nonrenewable sources of energy. Moreover, the romantic view of the traditional farm as a place where food is grown without synthetic chemicals remains a powerful anchor for the advocacy work of numerous groups that refuse to accept the premise that the tradeoff for modernization is the abandonment of chemically free agriculture.

Biotechnology, however, raises the possibility of an industrial-agricultural inversion, which has two characteristics. First, the farm will increasingly be used to manufacture products traditionally produced in industrial settings. Second, food production will take place to a greater degree outside of the farm, in enclosed continuous-process bioreactors. In "Food Without Farms" Walter Truitt Anderson describes a process of food production involving plant tissues, enzymes, and a basic nutrient feedstock.[32] He also describes research in which citrus juice vesicles are produced from cells in culture without need for the orange, grapefruit, or lemon. The change of food production from a land-based system to a tissue-culture system would give food producers much greater control over the output. First, food production would not be seasonally dependent, since tissue cultures can be grown in controlled industrial settings in a continuous process. This change promises the ultimate liberation of food production from abiotic and biotic factors affecting food security, including droughts, floods, crop diseases, and cutoff of fossil fuel supplies. The feedstock in the tissue-culture production system would still be agriculturally derived, and would still require some land-based farming. But crops grown only for use as feedstock would be much less vulnerable to environmental threats. "If the basic crops were trees or brush, they would require less fertilizer and water, and they would probably need less expended effort to help them fight off their enemies." So instead of farms as we know them, with their seasonal crops, we would have plantations of trees or brush that would be harvested periodically as needed. The wood would then be broken down by a biochemical process into simple sugar syrups, eventually transported to food factories.[33]

The trend in agricultural biotechnology is to allocate land-based agriculture for the raw materials of production. Differentiation of food products, once controlled by farmers, may begin to shift to industrial tissue-culture farms, just as food processing gained prevalence after the World War II when improved refrigeration and transport technologies were employed.

Proteins harvested from bacteria (called single cell proteins) are already being used as a source of animal feed,[34] as microorganisms convert an inexpensive waste feedstock into protein. Jeffcoat cites as possible feedstocks carbon dioxide, methane, methanol, sugars, and various hydrocarbons.[35] Even if the feedstock for microbial fermentation derives from land-based agriculture, the new metaphor of the farm may more closely approximate a mining industry in which raw materials (carbohydrates and sugars) are "mined" and subsequently processed into consumer products (proteins).

Traditional boundaries between agriculture and industry, particularly in drug and chemical manufacture, are becoming blurred. Transgenic animals have been modified with foreign genes that express scarce human proteins. Through this process, called "gene pharming," useful pharmaceutical products have been produced in the milk of sheep, goats, and cows. A Dutch company called Gene Pharming Europe BV genetically engineered the world's first bull with a gene that allows his daughters to produce human milk protein.[36] This process is beneficial because the cost of producing drugs in animal milk is only a fraction of the cost of producing them in a traditional bioreactor. Gene pharming typifies the role reversal that occurs when industrial production is transferred to the farm.

At another blurred boundary between agriculture and industry, genetically modified tobacco plants are being used to produce enzymes for the food industry. Polyhydroxybutyrate (PHB) is a biodegradable thermoplastic derived from many species of bacteria. The genes from a bacterium that encodes the enzymes required for synthesizing PHB were placed in a tobacco plant virus, which then may infect a plant by inserting the desired genes. If this promising technique works, it may be the initial step toward producing novel biopolymers in plants through genetic engineering.[37]

Although biotechnology calls into question the distinction between products of industrial and agricultural origin, the blurring of that distinction has considerable precedent. The modern high-efficiency farm uses Tayloristic assembly-line practices. The machinery must be supplied from sources outside of the farm's operation. And the Industrial Revolution created a new demand for selected agricultural crops such as cotton, wool, jute, rubber, and vegetable oils.[38] Food crops themselves may undergo processing or treatment prior to distribution in consumer markets. Innovations in biotechnology are thus emphasizing an agricultural-industrial nexus that has evolved for more than a century.

The distinction between rural and urban contributions to an industrialized society can no longer be grounded in an essentialist view of production sector contributions to the economy, lifestyles, cultural values, or production modalities. Food, fibers, and commercial chemicals may come either from an industrial processing plant or from land-based agriculture. If biotechnology is successful, one analysis predicts that

> the elimination of major portions of the farming enterprise [will] displace farmers and farmworkers on a scale never before possible. . . . [W]e can expect the not-so-gradual reduction of spatial, temporal, and climatic barriers to food and fiber production. This change alone will bring with it substantial social upheavals as the location of production changes. In addition, we can expect the elimination of major portions of the farming enterprise if field crops are grown in vitro."[39]

These changes notwithstanding, the concept of the traditional small family farm rooted in a popular mythology persists, possibly because of its significance in the nation's history (over a million small farms still operate in the United States) and secondarily because the farm is an archetypal symbol of family values and independent entrepreneurship that people refuse to give up easily.

Conclusion: Genetic Power and Techno-Mythmaking

Every ancient culture has created myths. They are the stories in which fantasy, reality, moral education, sacred values, cosmogony, life's lessons, and the passing on of valued traditions are mixed together. In the many studies of myths in history, scant attention has been given to mythmaking in contemporary society. Perhaps the arrogance of scientific secularism prohibits science from seeing itself as creating myths, an activity it views as relegated to prescientific societies and studies of folklorists. Mythmaking, after all, is a revelatory approach to truth and a substitute for scientific rationality.

But if we view mythmaking as a process of inventing powerful new symbols that introduce a set of values or expectations, that explain our origins or our essential being, and that help define a path to the future—then it is possible to see the roles that myths have held in contemporary societies.

Cultural anthropologists have learned that myths are more than expressions of primitive wisdom. Some have observed that myths are the precursors to a system of law and moral truths in that they provide a unifying cognitive structure for social cohesion. Myth fulfills an indispensable function in primitive culture: It expresses, enhances, and codifies belief; it safeguards and enforces morality; it vouches for the efficiency of ritual and contains practical rules for the guidance of human actions.[40]

Mythmaking in contemporary science has less to do with practical rules for the guidance of individual human behavior than it does with the choice of future pathways for societal progress. Techno-myths, as I call them, provide hopeful symbols and comforting beliefs during periods of uncertainty, anxiety, and change. These beliefs may be speculations, exaggerations, or even false notions of hope, but they are designed to achieve social commitment to unanimity of purpose. Scientific myths are anchors of belief in a selected conception of modernity. "Myth is above all a cultural force . . . an indispensable ingredient of all culture."[41] For example, each technological revolution has its power myth. Nuclear energy was going to produce so much inexpensive electricity we would not have to meter it. The myth of nuclear energy was replaced with the myth of fusion power, according to which energy would be not only plentiful but also compatible with a safe environment. We witnessed the myth of DDT, a safe and universal solution for the eradication of insect pests, followed by a grander myth about a chemical utopia built on synthetic organic molecules.

After decades of studying myths in native cultures, anthropologists began to recognize the parallels in functional roles between ancient and modern myths. Mythmaking is ultimately about securing beliefs.

Every culture creates and values its own myths, not because it may not be able to distinguish between truth and falsity but because the function of myths is to maintain and preserve a culture against disruption and destruction. Myths serve to keep people struggling against defeat, frustration, and disappointment; and they preserve institutions and institutional processes.[42]

A generally, although perhaps not universally, accepted tenet is that genetic technology is capable of transforming biological entities significantly beyond what could be achieved with traditional technology. Modern biotechnology has brought biology from a predominantly analytical phase to a new synthetic phase in its historical evolution. The possibilities for rearranging species are for all practical purposes unlimited. Various interests are involved in a struggle over the power images of modern biology. Power is central to current mythmaking. For staunch supporters of biotechnology, genetic power translates into investor confidence. Venture capitalists ride the wave of biotechnological power while the established corporate sectors reinforce the notion that genes hold the key to a new economic order.

In the postchemical phase of industrialization a new folklore of a healing technology has emerged. According to this new wisdom biotechnology will reestablish our balance with nature; it will offer a cornucopia of curative and safe products. The same promoters of biotechnology are not blind to the critics who also embrace the metaphor of genetic power. In this case, however, power is read as the potential for a technological disaster. For this reason power must be played down, especially in communications with the media. The abundant discourse on biotechnology cites its power to cross species barriers and to introduce completely new traits into indigenous organisms.[43] Periodically, scientists dispute this message, as a *Boston Globe* headline illustrates "Splice Genes? Nature Did It Long Before Geneticists."[44] There is no dearth of inconsistency in how the science media report on the power of genetic technology.

Promoters of biotechnology must walk a fine line in this game of symbols. They accept the power metaphor but interpret it as inherently safe power, like the power of solar energy. The grand techno-myth is that with genetic engineering we can fine-tune nature, preserve its diversity while reaping its bounty. Organisms can be genetically modified "in less time and with greater precision, predictability and control than possible with traditional methods."[45]

In the cultural sphere of environmental activism, the use of the power metaphor evokes apprehension. Political activists exploit the power metaphor by emphasizing risk and uncertainty. There is no safe power. The means by which one controls a powerful technology is not with more technology. Therefore, unless the social values are worked out in advance, genetic power is greeted with suspicion. While critics emphasize genetic power as a threat, they also dismiss it as a false power that will not solve societal problems.[46] In the social arena, stakeholders who may disagree about the value of biotechnology continue to advance the myth of genetic power. Not only has the mythology developed around the power of techniques, according to a recent study by Ruth Hubbard and Elijah Wald, the gene sui generis has been afforded a special metaphysical status. "The myth of the all-powerful gene is based on flawed science that discounts the environmental context in which we and our genes exist."[47] Ultimately, as the myth of genetic power gains in significance, less emphasis is placed on alternative technologies and on social determination as opposed to corporate determination of technological futures. Myths are social constructions designed to protect beliefs and invoke order. If this interpretation is correct, then the strongest effort at mythmaking for biotechnology is expected in those periods where the greatest threat to the success of commercialization of biotechnology takes root, for example, periods of social mobilization against specific products or technologies.

Once embedded in public consciousness, techno-myths are difficult to displace. For this reason different constituent groups view as essential the struggle to control the images of biotechnology at the outset. Successful techno-myths will blunt society's critical perspective. New information inconsistent with the orthodox view is easily discredited or ignored. An independent intelligentsia has the responsibility of comparing the constructed images of biotechnology with empirical reality. If that role is relinquished, biotechnology will become self-reifying. Neither the media, the general public, nor much of science will be in a position to distinguish appearance from reality.

Versions of this essay have appeared in *Issues in Agricultural Bioethics*, ed. T. B. Mepham, G. A. Tucker, and J. Wiseman (Nottingham: Nottingham University Press, 1995), chapter 1; and Sheldon Krimsky and Roger Wrubel, *Agricultural Biotechnology and the Environment* (Champaign-Urbana: University of Illinois Press, 1996), chapter 11.

Notes

1. Raymond A. Zilinskas and Burke K. Zimmerman, eds., *The Gene-Splicing Wars* (New York: Macmillan, 1986).
2. The following propositions that contribute to the mythmaking potential of biotechnology will not be addressed in this paper: Biotechnology will (will not) feed the world's hungry people; biotechnology will (will not) provide cures for the world's major diseases; biotechnology will (will not) lead us to renewable resource economies. These claims have also been used to advance the biotechnology research agenda and to implant the idea in the minds of the general public that biotechnology is synonymous with progress.
3. Gregg Easterbrook, *A Moment on the Earth: The Coming Age of Environmental Optimism* (New York: Viking, 1995), p. 145.
4. Bruce N. Ames and Lois S. Gold, "Too Many Rodent Carcinogens: Mitogenesis Increases Mutagenesis," *Science* 249 (1990), 970–71.
5. U.S. Food and Drug Administration, Department of Health and Human Services, "Statement of Policy: Foods Derived from New Plant Varieties," *Federal Register* 57 (1992), 22984–23005.
6. Charles S. Gasser and Robert T. Fraley, "Transgenic Crops," *Scientific American* 266 (1992), 62–69.
7. Susan Harlander, "Social, Moral and Ethical Issues in Food Biotechnology," *Food Technology* 45, 5 (May 1991), 152; and Melanie Miller, "The Promise of Biotechnology," *Journal of Environmental Health* 54, 2 (Sept.–Oct. 1991), 13–14.
8. Susan Harlander, "Food Biotechnology: Yesterday, Today, and Tomorrow," *Food Technology* 43, 9 (Sept. 1989), 196–206, on p. 200.
9. National Academy of Sciences, Committee on the Introduction of Genetically Engineered Organisms into the Environment, *Introduction of Recombinant-DNA Engineered Organisms into the Environment: Key Issues* (Washington, D.C.: National Academy Press, 1987), on p. 21; and Sheldon Krimsky, *Biotechnics and Society: The Rise of Industrial Genetics* (New York: Praeger, 1991), p. 142.
10. Krimsky, *Biotechnics and Society* (note 9), p. 142.
11. Rebecca Goldburg, Jane Rissler, Hope Shand, and Chuck Hassebrook, *Biotechnology's Bitter Harvest*, A Report of the Biotechnology Working Group (Washington, D.C.: National Wildlife Federation, 1990), p. 55. A study by the Union of Concerned Scientists supported these arguments and concluded that transgenic crops pose serious environmental risks; see Jane Rissler and Margaret Mellon, *Perils Amidst the Promise* (Cambridge, Mass.: Union of Concerned Scientists, 1993).
12. Dale A. Miller, "Environmentalism and the New Agriculture," speech delivered at the Ag Bankers Association Annual Meeting, Denver, Colorado, 12 Nov. 1990, in *Vital Speeches of the Day*, 15 January 1991, pp. 221–24; and Earie H. Harbison, Jr.,

"Biotechnology Commercialization—An Economic Engine for Future Agricultural and Industrial Growth," speech delivered to the World Economic Forum, Davos, Switzerland, 5 Feb. 1990.

13. Dick Russell, "Miracle or Myth?" *Amicus Journal* 15, 1 (1993), 20–24; and Pamela Weintraub, "The Coming of the High-Tech Harvest," *Audubon* 94, 4 (1992), 92–103.

14. John H. Perkins, "The Quest for Innovation in Agricultural Entomology, 1945–1978," in *Pest Control: Cultural and Environmental Aspects*, ed. David Pimentel and Perkins (Boulder, Colo.: Westview Press, 1980), pp. 23–80.

15. Martin H. Rogoff and Stephen L. Rawlins, "Food Security: A Technological Alternative," *BioScience* 37 (1990), 800–807.

16. Les Levidow, "Codes, Commodities, and Combat: Agricultural Biotechnology as Clean Surgical Strike," in *Perspective on the Environment II*, ed. S. Elworthy et al. (Aldershot, UK: Avebury, 1995).

17. David Pimentel et al., "Environmental and Economic Cost of Pesticide Use," *BioScience* 42 (1992), 750–60, on p. 750.

18. Jozef Schell, Bruno Gronenborn, and Robert T. Fraley, "Improving Crop Plants by the Introduction of Isolated Genes," in *A Revolution in Biotechnology*, ed. Jean L. Marx (Cambridge: Cambridge University Press, 1989), pp. 130–44.

19. George P. Georghiou, "Implications of Potential Resistance to Biopesticides," in *Biotechnology, Biological Pesticides and Novel Plant-Pest Resistance for Insect Pest Management*, ed. D. W. Roberts and R. R. Granados (Ithaca, N.Y.: Boyce Thompson Institute for Plant Research, Cornell University, 1988), p. 137.

20. Fred Gould, "Ecological-Genetic Approaches for the Design of Genetically Engineered Crops," in *Biotechnology*, ed. Roberts and Granados (note 19), pp. 146–51, on p. 146.

21. Harlander, "Food Biotechnology" (note 7), p. 196.

22. Dan L. Burk, Kenneth Barovsky, and Gladys H. Monroy, "Biodiversity and Biotechnology," *Science* 260 (1993), 1900–1901, on p. 1901.

23. Jack Doyle, *Altered Harvest* (New York: Viking, 1985).

24. Michael Fox, *Superpigs and Wondercorn: The Brave New World of Biotechnology and Where It All May Lead* (New York: Lyons and Burford, 1992), p. 133.

25. Henk Hobbelink, "Plant Patents: Who Benefits?" *GeneWatch* 4, 6 (June 1987), 6–7, on p. 8.

26. Vandana Shiva, "Biodiversity, Biotechnology and Profit: The Need for a People's Plan to Protect Biological Diversity," *Ecologist* 20, 2 (March–April 1990), 44–47; and Shiva, *Monocultures of the Mind: Biodiversity, Biotechnology and Scientific Agriculture* (London: Zed, 1993).

27. William K. Stevens, "Are Gene-Altered Plants an Ecological Threat? Test Is Devised" *New York Times*, 22 June 1993, p. C4; and Roger P. Wrubel, Sheldon Krimsky, and R. E. Wetzler, "Field Testing Transgenic Plants," *BioScience* 42 (1992), 280–89.

28. Gus Speth, "A Luddite Recants," *Amicus Journal* 11, 2 (1989), 3.

29. Darryl R. J. Macer, *Shaping Genes* (Christchurch, N.Z.: Eubios Ethics Institute, 1990), p. 32.

30. Bernard Dixon, "Biotech's Effects on Biodiversity Debated," *Biotechnology* 8 (1990), 499.

31. Harlander, "Food Biotechnology" (note. 7), p. 197.

32. Walter Truett Anderson, "Food Without Farms: The Biotech Revolution in Agriculture," *Futurist* 24 (1990), 16–22.

33. Ibid., p. 20.

34. S. R. L. Smith, "Single Cell Protein," *Philosophical Transactions of the Royal Society* B290 (1980), 341–54.

35. Roger Jeffcoat, "The Impact of Biotechnology on the Food Industry," in *Biotechnology: The Science and the Business*, ed. Vivian Moses and Ronald E. Cape (London: Harwood Academic Publishers, 1991), pp. 463–80, on p. 478.

36. "Herman, the Biotech Bull, May Sire New Drug Era," *Journal of Commerce*, 29 Dec. 1992.

37. Y. Poirier, D. E. Dennis, K. Klomparens, and C. Somerville, "Polyhydroxybutyrate, a Biodegradable Thermoplastic, Produced in Transgenic Plants," *Science* 256 (1992), 520–23.

38. David B. Grigg, *The Agricultural Systems of the World: An Evolutionary Approach* (Cambridge: Cambridge University Press, 1974), p. 284.

39. Lawrence William Busch, William B. Lacy, J. Burkhardt, and L. R. Lacy, eds., *Plants, Power and Profit: Social, Economic, and Ethical Consequences of the New Biotechnologies* (Cambridge, Mass.: Blackwell, 1991), p. 190.

40. Mircea Eliade, *Myth and Reality*, trans. Willard R. Trask (New York: Harper, 1963), p. 20.

41. Bronislaw Malinowski, *Myth in Primitive Psychology* (New York: W. W. Norton, 1926), pp. 91–92.

42. R. Gotesky, "The Nature of Myth and Society," *American Anthropologist* 54 (1952). Also cited in Gregor Sebba, "Symbol in Modern Rationalistic Societies," in *Truth, Myth and Symbol*, ed. Thomas J. J. Altizer, William A. Beardslee, and J. Harvey Young (Englewood Cliffs, N.J.: Prentice-Hall, 1962), p. 141.

43. Steve Olson, *Biotechnology: An Industry Comes of Age* (Washington, D.C.: National Academy Press, 1986), p. 3.

44. Bruce Fellman, "Splice Genes? Nature Did It Long Before Geneticists," *Boston Globe*, 6 April 1992, p. 45.

45. Susan Harlander, "Biotechnology: A Means for Improving Our Food Supply," *Food Technology* 45, 4 (April 1991), 84–95, on p. 84.

46. Brian Tokar, "The False Promise of Biotechnology," *Z Magazine*, Feb. 1992, p. 27.

47. Ruth Hubbard and Elijah Wald, *Exploding the Gene Myth: How Genetic Information Is Produced and Manipulated by Scientists, Physicians, Employers, Insurance Companies, Educators, and Law Enforcers* (Boston: Beacon Press, 1993), p. 6.

Part III
The Molecular Workplace

Monoclonal Antibodies: From Local to Extended Networks

Alberto Cambrosio and Peter Keating

Antibody production by the immune system of an individual involves a network of cellular interactions with complex and precise regulatory circuits. A comparable network should be organized between the laboratories involved in the hybridoma technology in order to avoid wasting of manpower and financial support.[1]

Monoclonal antibodies (MAbs), once an esoteric technique utilized in a few laboratories, are now a tool with widespread applications in a number of disciplinary, clinical, and industrial domains. Effecting this transformation required the establishment of a complex infrastructure, one on which the ongoing existence and circulation of MAbs qua tools is still predicated. This article focuses on the investments underlying that process, and our analysis calls into question conventional distinctions between public science and private science.

Several scholars in science studies have looked with mixed feelings at the parallel development of biotechnology and molecular biology in recent years. They have often expressed the fear that the present trend toward privatization and commercialization in the life sciences will increasingly limit or even threaten the development of "public science," by which is meant a kind of science performed mainly in universities and other public arenas and characterized by the free circulation of ideas, results, and research materials.[2] The expression of these fears has not been confined to historians and sociologists of science but has also found an echo, or possibly its origin, among several (although not a majority of) scientists, politicians, and journalists.[3] These critics have taken the exponential growth of patents issued in the life sciences as evidence of a trend towards increased secrecy, conflict of interest, and the alteration of the traditional career path of scientists.[4] That transformation of the political economy of (bio)technoscience is thought to go beyond its "social" aspects to affect the actual content of fundamental research.[5] Thus the rise of biotechnology and venture capital is believed partly responsible for such endeavors as the "interferon crusade" described by Sandra Panem—that is, the promotion of interferon as an anti-cancer wonder drug by scientists, foundations, and industrialists, in spite of equivocal clinical results.[6]

Our aim here is neither to support nor to dispute the validity of this kind of claim, since we consider such a "sociology of accusation" to be problematic. Analysts who argue about the value or the danger of privatization adopt the dubious position of being part and parcel of a controversy while claiming to have found an Archimedian point. Moreover, they take for granted the categories used by the actors, thus eschewing investigation of the critical work the actors deploy in constructing the various "sides" of a controversy—the "endogenous critical inquiry," as Michael Lynch terms it.[7] These authors seem to share, for instance, the actors' belief in the *inherently public* nature of the scientific enterprise. They consequently depict privatization as a form of "hijacking" of public goods by private interests.

The focus on privatization as the active process obscures a question of utmost interest to a sociology of (biomedical) sciences: namely, how is public science, qua public science, constructed? As Michel Callon cogently argues, conventional assumptions about the public status of science—that science is by its nature a public good, that no effort or particular investment is needed for its results and products to circulate freely—contradict the results of ethnographically oriented inquiries into scientific practice.[8] Public science is made possible by investments and interventions of unsuspected breadth and width. A lot of work is needed to make science public; (almost) no work is needed to keep it private. Laboratory studies and the analysis of scientific networks have replaced the belief in some form of magic correspondence between laboratory results and "nature out there" with a detailed analysis of the various processes involved in "making things work" and in constructing their generality—both in the epistemic sense of generally valid results and in the material sense of being able to reproduce and implement results in laboratories and clinics scattered around the (mostly Western) world.[9]

An obvious example of these activities is the establishment and enforcement of standards and metrological conventions to ensure that laboratories around the world use procedures and material that are "for all practical purposes" identical, thus creating the basic conditions for the successful pursuit of technoscience.[10] And if the focus on material practices seems ill-suited to an account of the circulation of "abstract" ideas, then one should consider not only that those ideas can hardly be separated from the instruments and practices that translate them into meaningful operational statements, but also that an enormous investment in teaching and training is necessary to create a group of people to whom those statements do indeed make sense.[11] We would argue, then, that it is only our prereflexive acquaintance with the modern scientific infrastructure that blinds us to its existence and induces us to accept as unproblematic any claim about the free circulation of scientific results.

Callon's claim does not merely invert the conventional line of argument about public science, from "What is involved in subtracting something from the public domain?" to "What is involved in making something public?" More decisively, he questions the analytic primacy of the public-private dichotomy, suggesting that it is fruitfully replaced by a distinction between local and extended networks. Local science is always private, in the sense that it does not circulate and it thus remains in private hands. Local networks must be transformed into or linked to extended networks to create the conditions of private appropriation. If such appropriation is to appear financially worthwhile, the objects being privatized must be able to circulate widely.

In this article, we shall exemplify this argument by looking at one particular innovation in the field of recent biological and biomedical sciences: hybridoma technology (HT), used to produce so-called monoclonal antibodies (MAbs). While possibly less fashionable as a topic of social scientific inquiry than genetic engineering, HT-related activities, which gave biotechnology its first long-awaited marketable products, seem prima facie to constitute a perfect example of the privatization trend.

MAbs as an Example of Privatization

According to the canonical version of events, HT was developed in Britain in 1975 in a fundamental research setting—the Molecular Biology Laboratory of the Medical Research Council in Cambridge—by a leading immunogeneticist, César Milstein, and a postdoctoral fellow, Georges Köhler. They jointly received the 1984 Nobel Prize in medicine for this achievement, which is credited with having "revolutionized serology, provided tools for the examination of basic immunological mechanisms, and created an industry."[12] Indeed, the technique provided the necessary cash flow for hundreds of start-up biotech companies, who sold MAbs as research reagents and as part of diagnostic kits. Milstein and Köhler did not seek legal protection for their achievement, but several researchers and companies (especially in the United States) soon engaged in successful attempts to patent various aspects of HT.

We shall briefly recall three major episodes. A bitter polemic developed between British and American researchers when a team from the Wistar Institute in Philadelphia applied for patents covering what, in Milstein's (and several other scientists') eyes, was a straightforward application of the original technique developed in the Cambridge laboratory to two areas of obvious medical and commercial interest—tumor and viral antigens.

Wistar's researchers held stock in and consulted for Centocor, a biotech startup company in the HT field.[13] The patents were issued in the United States but rejected in the United Kingdom, precisely on the ground of obviousness.[14] To our knowledge, the Wistar patents were never tested before the courts, and although some institutions did ask for and obtain licenses, many others did not bother. From this point of view, Wistar's patents are hardly comparable to the patent that Stanley N. Cohen and Herbert W. Boyer took out in the genetic engineering.

Two subsequent episodes are generally thought more consequential for the definition of HT and its products as public or private goods. The first was a dispute between two biotechnology start-up companies formed specifically to exploit HT, Hybritech and Monoclonal Antibodies Inc., over a patent covering the use of MAbs in conjunction with a specific immunodiagnostic technique (so-called "sandwich immunoassays")—once again, a direction several scientists viewed as obvious. Hybritech won its case against Monoclonal Antibodies, then sued a giant in the immunodiagnostic field, Abbott Laboratories, which settled out of court.[15] In the second episode two large biomedical companies, Becton-Dickinson and Ortho Pharmaceutical Corporation (a subsidiary of Johnson & Johnson), fought over patents covering MAbs against human lymphocyte subpopulations and their use in conjunction with sophisticated equipment, the so-called flow cytometers. These MAbs are used, among other things, in the diagnosis of AIDS and leukemias and in monitoring and treating transplant patients, and several researchers from different laboratories claimed to have independently generated them. Becton-Dickinson settled out of court; Ortho subsequently sued Coulter, which also settled out of court.[16] In both cases, considerable commercial interests were at stake. Given the absence of a strong patent covering the basic technique, some perceived these two sets of patents as the most effective strategy for appropriating the field. One immunoassay specialist noted that "by successfully patenting the use of antibodies produced by Milstein and Köhler's technique, commercial companies have succeeded in implicitly patenting the invention itself."[17]

In 1990 we analyzed these patent disputes as indicators that the domain of public science was progressively being privatized, arguing that they hinted at "a subtle but significant realignment in the political economy of science and technology." Driven by economic forces, this realignment took place at the interface between free information—that is, all the information that can be appropriated without legal or financial restraints, which constitutes the common pool of knowledge accessible, in principle, to everybody—and proprietary information—information protected by law and taking the form of "know-how" or patents. Indeed, one ground on which the validity of both patents was contested was that what was being patented—transformed into proprietary information—was in fact free scientific information. The very existence of these disputes led us (and most certainly several scientists we interviewed) to conclude that in the new biotechnological context, "when used in arbitration between [free information] and [proprietary information], [free information] has no voice for itself but only against itself." Thus we argued that in the aftermath of these patent disputes, scientists who wanted "to protect their conceptions from being appropriated as proprietary information may have to withhold their ideas from the public domain until they are fully formed and can be thoroughly articulated as, for example, 'obvious' rather than 'obvious to try.' . . . [In addition,] open scientific research would have to adopt the recording practices of commercial research to defend its product *as* free scientific information."[18]

Our argument certainly captured the fears (or the hopes) of some scientists. It was silent, however, on the amount of work that went into creating the conditions that allowed MAbs not only to circulate freely, but also to be translated into a generic tool that many felt should circulate freely. An alternative account of the events that led to the patent disputes described in our 1990 paper, then, would be based on the following three arguments. (1) MAbs were not, at the outset, generic *tools* destined for worldwide application: they had to be translated into tools and to be made general. (2) This translation involved a movement from local networks, where HT had first found a place, to extended networks. (3) This movement was, in turn, grounded in the establishment of a complex infrastructure, upon which the ongoing existence and circulation of MAbs qua generic tools is still predicated.

The Translation of MAbs into Tools

Let us first quickly sketch the first point. The 1975 *Nature* paper by Köhler and Milstein ended with a paragraph hinting at the possible generic value of the technique—that hybrid antibody-producing "cultures could be valuable for medical and industrial use." But the only results the authors actually reported in 1975 concerned the production of antibodies against sheep red blood cells, and they acknowledged that "it remains to be seen whether similar results can be obtained using other antigens"; they did not specify the exact meaning of the term "results." To be sure, the 1975 paper did not mention "hybridoma technology," and the original purpose of the research was clearly not to develop such a technology.[19] The paper was at first viewed as one among a number that used cell fusion techniques to dissect the genetic control of antibody diversity, the "evidential context" in which Milstein and Köhler's work was grounded.[20] But, at some point around 1977, the production of MAbs became an end in itself, no longer restricted to the initial immunogenetic network consisting of a few laboratories. This began a process leading to the establishment of the "generality" of HT and, correlatively, to the translation of MAbs into "a powerful new tool in biology and medicine."[21]

Milstein himself pointed out on various occasions how important focused interventions were for establishing the generality of MAbs as tools. In his Nobel lecture, for instance, he noted that he had for several years "shelved the antibody diversity problem to demonstrate the practical importance of MAbs in other areas of basic research and in clinical diagnosis."[22] In practice, this had meant teaming up with researchers from other fields to produce MAbs relevant to those domains. As a consequence, Milstein was able to build the generality of the technique *progressively*, showing that it could be applied in fields ranging from the clinical classification of leukemias, to transplantation, blood-typing, neurobiology, and large-scale protein purification. Of course, the future of MAbs did not lie with Milstein alone; other researchers and institutions were involved.

The preceding analysis makes it clear that the development of HT was not immediately recognized as an advance worthy of general acclaim. One leading immunologist, Melvin Cohn, commented: "Even after the Milstein paper appeared, most people wanted to use it for somatic cell genetics. We used [HT] mostly for that. We did not use it to make reagents. I don't think we really appreciated how important that part would be technically." Another researcher, Timothy Springer, wrote: "Soon after the publication of the hybridoma technique in 1975, it was not widely appreciated what a dramatic impact it would have. Widespread use of monoclonal antibodies came first in the field where they were developed, immunology."[23] Yet it could be argued that after a short period of delay, MAbs underwent a rapid diffusion process, and that therefore, in spite of initial hesitations or misunderstandings, this process was essentially unproblematic. The exponential growth of HT-related publications, depicted in Figure 1, seems to support such an interpretation.

A close look at *how* MAbs were transformed into tools shows that in fact they were first adopted as part of local networks—and there they replaced closely related elements, such as conventional antibodies. They only later appeared in extended rather than local networks. In other words, MAbs first appeared to be a substitute, and not a novelty. This is clearly why the use of MAbs is often described as obvious. The experimental system in use in local laboratories can be understood as a network of problems, statements, skills, equipment, and tools,[24] and it is precisely this network structure that creates a sense of obviousness when one of its components is replaced by another.

A few examples can be briefly sketched. At the Wistar Institute, then under the direction of Hilary Koprowski, Walter Gerhard produced one of the very first MAbs outside Milstein's laboratory, targeted against the influenza virus. Gerhard was working on that virus's antigenic variability, and to that end he had already used "monoclonal antibodies" that could be produced in small quantities and for limited periods of time with a technique developed by Norman Klinman.[25] Everything was therefore already in place for MAbs to function as part of the system: Gerhard knew not only how to

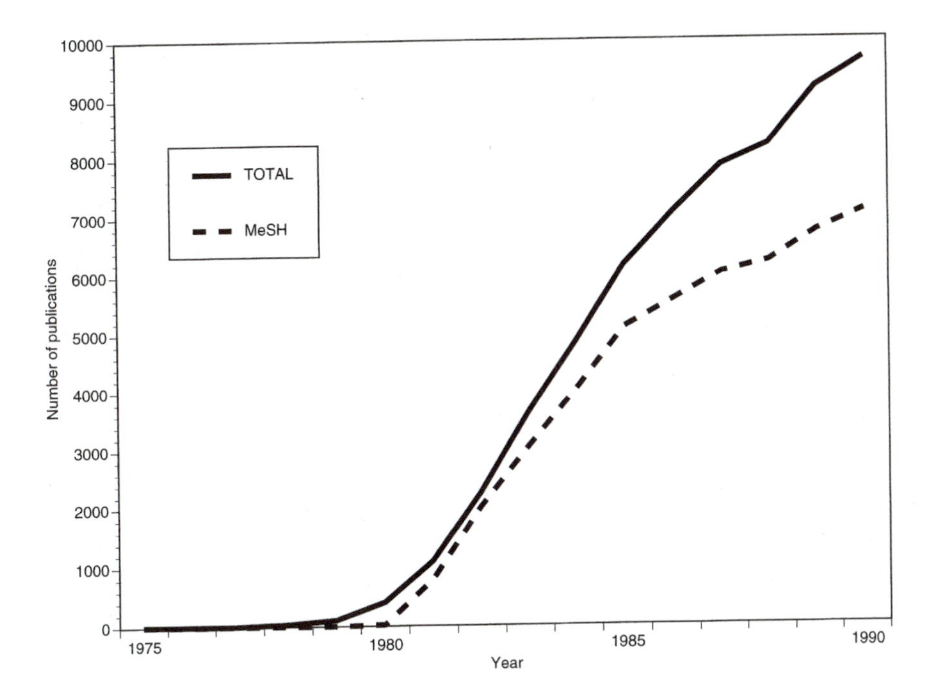

Figure 1. Number of publications indexed with the terms "monoclonal antibodies" and "hybridomas" in MEDLINE, the on-line version of *Index Medicus*, 1975–90. The terms were added to the list of official descriptors (Medical Subject Headings, MeSH) in 1982. The "total" line combines the number of publications indexed by the two MeSH terms, indicating that MAbs played a central role in them, and the number of publications where the terms are present in titles and abstracts but not as a MeSH, indicating a secondary role for MAbs. Data are shown for the pre-1982 period because MEDLINE staff unsystematically and retrospectively re-indexed a number of articles with the two descriptors.

deal with the technical aspects of their production but, more important, how to make sense of the new tool. To quote Gerhard: "Many people, they may have said, [HT] is nice but so what, what can we do with it? I didn't have to say that."[26]

Another early product of the Wistar Institute, anti-rabies MAbs, present a similar case. As William Wunner recalls, the Wistar people, including Koprowski and Tadeusz Wiktor, had studied the rabies virus for almost twenty years before the advent of MAbs. Thus once MAbs became available they were easily integrated into a preexisting experimental system: "What monoclonal antibodies have allowed us to do, we have always wanted to do. It has simply helped accomplish the goal."[27]

The immunoassay field provides another example. In simple terms, the technique of choice circa 1975 was labeled-antigen, rather than labeled-antibody, assays. Among those who advocated the labeled-antibody approach was the British scientist Roger P. Ekins. Ekins had participated in theoretical debates as to the technique's viability and possible advantages, but the impracticality of purifying the large quantities of conventional antibody needed had prevented him from implementing it. The various elements of the labeled-antibody assay package, including theoretical justifications, were thus already in place, and the next step was to replace conventional antibodies with MAbs. Not surprisingly, Ekins was among the first scientists to visit Milstein's Cambridge laboratory to inquire about how to produce the appropriate MAbs.[28]

Finally, there is the case of the apparatus known as the fluorescent activated cell sorter (FACS), developed by Leonard Herzenberg, a Stanford immunologist, in the late 1960s to early 1970s. Although the FACS later enjoyed a successful career in fundamental and clinical immunology, it originally had a problem: most of its applications relied on antibodies, and the variability of traditional serological reagents hampered comparisons between laboratories. MAbs promised to standardize the reagents and thus the laboratories, making the data they produced comparable. Herzenberg quickly adopted HT and became a leading advocate of MAbs. Once again, MAbs happened to fit into a preestablished package: "I wrote a paper called 'Monoclonal antibodies and the fluorescence-activated cell sorter: complementary tools in cell biology.' I gave a lot of seminars with that title. Because they were very complementary tools. It was just perfect."[29]

Researchers could and did also take advantage of the opportunities offered by the new tool to develop new research directions—or, to use Callon's terminology, to reconfigure local networks.[30] In fact, we should qualify our previous statements by arguing that there is no such thing as "mere substitution." Replacing one item in an experimental system modifies that system and often explicitly redefines the balance between the "technical" and the "epistemic" elements of the system. For a new technique to be perceived as a bona fide tool, however, and for the opportunities it offers to become apparent, the technique must first be domesticated within an existing experimental system. These contingencies explain why MAbs were first applied to immunology and virology and only later reached areas where conventional antibodies had met with only limited success, such as bacteriology and parasitology. In bacteriology, for instance, publications using MAbs began to increase noticeably only in 1983.[31]

The transformation of a technoscientific object into a tool, while in practice a "messy" event, can thus be divided into two analytically distinct processes. The first is the use of that object as a technique within a given experimental system: a technique, in that sense, can only be of local interest, an idiosyncratic object among other idiosyncratic objects. The second process is the transformation of a local technique into a generic tool, one that involves a more stable reification, in which characteristics are attributed to the newly defined "tool" and stabilized, and its use in different settings is standardized. At this stage, the advantages of the new tool have to be assessed in generic terms and represented, precisely, as "generic advantages." This process does not free the tool from its connections to other relevant elements; it simply provides more extended networks within which the tool circulates.

Constructing Extended Networks

It is often claimed that the success of MAbs was linked to their standardizing power. The focus on standardization is a central and recurring argument in a programmatic document published by the NIH's National Institute of Allergy and Infectious Diseases in 1981, under the title *New Initiatives in Immunology.*[32] In its opening statements, the NIAID document ranked MAbs among "techniques and tools" (as opposed to factual findings) and (re)presented them as "reference reagents" that could be produced in "virtually unlimited amounts for indefinite periods of time." This capability would make them "available to all investigators or diagnostic laboratories needing them," and they could serve "as a reference reagent for as long as it is required." Past is the era, claimed the NIAID document, when the unpredictability and heterogeneity of the immune response made the production of antibodies more an art than a science. Now antibodies could be produced that share the standardized properties of chemical reagents, allowing results obtained by laboratories around the world to be standardized. It is difficult to resist the conclusion that it was not because MAbs were such a powerful tool that they led to standardization, but, rather, that they were translated into such a powerful tool precisely because they were seen from the very outset as a potential vehicle for standardization.

The production of MAbs as a standardizing tool obviously depended on the existence of a standardized biomedical infrastructure. Crises and accidents bring this often invisible infrastructure to light. For instance, a "massive mouse mixup" occurred in the United States in 1982, after a shipment of BALB/c mice used in the production of MAbs was found to be genetically contaminated. Various MAbs companies soon issued reassuring press releases indicating that the potentially disastrous consequences of this event had been avoided.[33] Breakdowns such as this one, independently of their outcomes, point to the work and arrangements necessary to provide scientists with a constant supply of standardized and thus "highly artificial" raw materials.[34]

The NIAID report advocated creating yet another, more specific layer of this biomedical infrastructure. First, MAbs were to be provided to inexperienced investigators unable to produce them; the NIH was therefore requested to encourage the "development of shared hybridoma facilities as part of a program of core grants." This provision would not be enough, however, since the uncontrolled proliferation of locally produced MAbs was certain to lead to confusion; consequently, steps would have to be taken "to identify those monoclonal antibodies that are most useful and to see that the same antibodies are used universally." To promote standardization and control the standardizing tools, the NIAID report also recommended that banking facilities be established and lists of available MAbs be published. But here again the use of "complex and individual terminologies" would have undermined the standardization effort. Thus the NIH was asked to encourage workshops and exchange of reagents, specifically in the area of "antibodies directed against subsets of human cells."[35]

These recommendations did not go unheeded. The call for workshops and exchange of reagents led to an international classification system for the surface antigens of human white blood cells, known as the "CD nomenclature." By 1986 the number of laboratories participating in the ongoing production of this nomenclature had risen to 190. They performed more than 150,000 assays to characterize about 800 antibodies. A precondition for present-day classification of lymphocytes is thus the network constituted to establish identity between MAbs produced by different laboratories.[36]

Several steps were taken to help prevent the anarchic proliferation of MAbs in various laboratories under idiosyncratic names. In 1983 a Hybridoma Data Bank (HDB) was established under the auspices of the Committee on Data for Science and Technology (CODATA) of the International Council of Scientific Unions and the International Union of Immunological Societies (IUIS). The HDB—which had three branches, located in the United States, Europe, and Japan—supplemented information available on an ad hoc basis from scientific publications and patents.[37] Just as necessary as providing information on the existence and availability of MAbs was establishing access to antibodies or to the antibody-producing cells, not merely making MAbs available to researchers but also regulating the circulation of cells and cell products. This thorny issue highlights the local and private nature of laboratory results and products, and the heterogeneous interventions needed to make them work outside their production site. As Lee Herzenberg recalls:

> Well, I can remember going to that meeting and announcing that we were making all these antibodies available and that anybody could have them and getting a tremendous positive response from the people at the meeting. There was almost a gasp when I said that all of these antibodies would be available. Then we got negative response from people who said that we had no right to make a unilateral decision because now everybody was forced to follow suit and that was true. Up until that time there were always discussions and decisions about [whether MAbs] should be made public.[38]

However, even the traditional mechanism whereby individual researchers send reagents to fellow scientists who request them was inefficient when viewed from the perspective of "worldwide standardization," a goal that Milstein and co-workers had already evoked in 1977.[39] This problem was solved by establishing centralized repositories to act as distribution centers. One such repository, which played a pioneering role, was created in a private research institution, the Salk Institute.

A bank of murine cell lines—cancer and immune-related—was first set up at the Salk Institute in the early 1970s. Operating under a National Cancer Institute contract, the immunologist Melvin Cohn then initiated a myeloma tumor program, using cell lines obtained mainly from Michael Potter at the NIH. Appropriately, this early investment in public infrastructure was one of the preconditions for Milstein and Köhler's work, since one of the essential ingredients of the 1975 Cambridge experiment was a tumor cell line that originally came from San Diego. More generally, other workers in the field have argued that Potter's and Cohn's tumor induction programs had an enormous impact: "Not only were the tumors themselves important, but their distribution promoted an interchange of ideas and reagents that has played a crucial role in the progress of modern immunology."[40]

The Salk Institute cell bank played a central role after 1975 as well. Applying to NCI for contract renewal in 1981, Cohn explained in the budget justification section that "this new contract period reflects for the first time the shift in emphasis toward hybridoma technology." Noting that "the volume of work has increased disproportionately with the number of cell lines maintained and shipped because of the increased number of hybridomas in the library," Cohn further argued that the cell bank was "becoming the unofficial standard reference center for hybridoma cell lines." The cell bank's role was not simply to make cell lines available, but also to provide researchers with essential, unpublished information coming "from other investigators who will call or write to our project manager," information that complemented the "materials and methods" section of published articles.[41]

The American Type Culture Collection (ATCC) soon took over the function fulfilled by the Salk Institute cell bank. In October 1980 the agency began operating the Hybridoma Cell Bank with support from NIAID, and ATCC's increased responsibilities in this field found a material counterpart in the construction of a new building at its headquarters in Rockville, Maryland. In 1981 ATCC personnel transferred the holdings of the Salk Institute cell bank to the newly created ATCC Tumor Immunology Bank. Commercial enterprises also, such as Becton-Dickinson and Ortho Pharmaceuticals, played an important role in this domain. By putting a large number of MAbs on the market, they not only ensured the circulation of those MAbs, but also created a standard. In several cases, individual researchers urged or helped companies to enter the business of selling MAbs. Herzenberg was among them:

> I think that [the fact of] the companies being involved and making these antibodies was actually very important to the development of the whole field because it was very difficult for one laboratory to develop a whole library or battery of reagents that it needs for its work. . . . The idea of hybridomas was that everybody should have the same specificity, reactivity, same amount of color, and reagents. So if everybody made their own, there was great difficulty logistically in everybody making their own. It also defeated the purpose, the great advantage of monoclonals. So I felt it should come from central sources. [Becton-Dickinson] agreed that I could train their people so they would be very good at making standard reagents and keep the quality up.[42]

Not all researchers—especially academic researchers—felt this way. For instance, after Patrick Kung first developed for Ortho the series of MAbs against human lymphocyte subpopulations mentioned above, he and his coworkers teamed up with a unit led by Stuart F. Schlossman at the Dana Farber Cancer Institute in Boston, in a collaborative effort to characterize them. The collaboration between the two groups was fraught with tensions, especially over how to control the distribution of the antibodies. The industry team wanted to make them available to all qualified investigators who requested them, precisely in order to build their generality, whereas the academic team would have preferred to restrict their initial use to a few private hands, in order to clean up in the field.[43] This attitude seems to have been widespread in the field, as Herzenberg recalled: "For the most part everybody was playing very close to the bat. They gave their friends some antibodies but they were not going to give the antibodies out. They were generally not going to make them publicly available."[44] Both in the Ortho case and in the case of the collaboration between Herzenberg and Becton-Dickinson, private ownership of the hybridomas became a key to public circulation.

It is apparent, then, that the issue pertains less to ownership per se, or to public researchers versus private companies, than to constructing an infrastructure that allows the circulation of reagents and thus specifies the extent to which and the circumstances under which specific technoscientific objects can be transferred from local to extended networks.

Conclusion: Public Science Versus Private Science

A meeting was held in 1981 to mark the dedication of the new ATCC building that housed the Tumor Immunology Bank, and at it a NIAID employee, Henry Krakauer, read a paper that asked a revealing question: "Why should the government use tax funds to support the quite expensive activities associated with [central banking facilities for hybridomas]?" His answer was that "hybridoma technology generates monoclonal antibodies of such a wide variety, representing specificities of various shadings and affinities for the many antigenic determinants on the molecules and cells, [that] any one laboratory can sample and investigate only a small fraction, and [that] very many laboratories are active in this field."[45] That situation was potentially chaotic, raising the specter of utmost confusion about which antibody was which and which antibody did what. Large investments (and central cell banks are only the tip of the iceberg) were thus needed not simply to make MAbs available to all laboratories, but, more important, to define and stabilize their properties.

Krakauer's presentation was accompanied by several others discussing patent issues, not unsurprisingly since the deposit of specimens of the biological material in specialized repositories (such as the ATCC) is one of the requirements for patenting hybridomas. Deposited cell lines must be made available to anyone on request. A hybridoma bank, according to Krakauer, "could not function as a relatively passive repository and intermediary, but would require an active quality control program as well as an investigational component that would familiarize the staff with the tricks and pitfalls of the field and thus permit them to act as an independent source of technical guidance."[46] This brings us back to our opening thesis: for hybridomas to be patentable, not only must an extended network be available to circulate and stabilize them, but "public" and "private" realms, scientific and legal activities must interact to produce this network.

Krakauer's presentation thus supports the main arguments of this article. The translation of MAbs into tools was coextensive with the construction of their generality, a process that involved defining their "advantages" and "disadvantages" relative to alternative tools and stabilizing their characteristics. In other words, it was not simply a question of providing channels that allowed the physical circulation of reagents and cells endowed with some particular properties. Rather, the existence, maintenance, and stabilization of those properties depended on the establishment of such a large network. As the reader will recall, worldwide standardization was seen as a major justification for supporting HT, and that standardization was made to rest on the definition of MAbs as "pure" substances. Immunology could turn into chemistry since antibodies could now be produced that shared the properties of chemical reagents. Yet, as Gaston Bachelard argued many years ago, we should not forget that "the purity of a substance is . . . the work of man. . . . It retains the essential relativity of human works." Purification processes reveal the industrial nature of modern science and presuppose the existence of a coherent set of socially agreed upon, mutually defining reagents.[47] What counts as "purity" at any given time is the result of the regulation of clinical, laboratory, and industrial practices.

Regulation presupposes circulation: in the present case, the extension of MAbs beyond the local networks where they were first confined. MAbs had to be taken out of the private hands of individual researchers to become what we today perceive them to be: generic tools of widespread application. In this sense, "private"—that is local—science performed in academic or industrial laboratories had to be transformed into "public" science, science conducted in an extended network that circulates both results and research materials. The terms "private" and "public," as used in this article, no longer refer to the issue of ownership but, rather, to the notion of circulation. This use of those terms might be somewhat counterintuitive, but it has the advantage of focusing our attention on the complex and often overlooked processes involved in constructing what is generally referred to as "public science."

Research for this article was made possible by Social Sciences and Humanities Research Council of Canada Grant 410-91-1935.

Notes

1. J. P. Revillard and J. Cohen, "Monoclonal Antibodies: An Overview of Their Advantages and Limitations in Nuclear Medicine," in *Nuclear Medicine and Biology Advances* (4 vols.), vol. 3, ed. Claude Raynaud (Oxford: Pergamon Press, 1983), pp. 2287–91, on p. 2290.

2. See, e.g., Edward J. Yoxen, *The Gene Business: Who Should Control Biotechnology?* (New York: Oxford University Press, 1985); Martin Kenney, *Biotechnology: The University-Industrial Complex* (New Haven, Conn.: Yale University Press, 1986); Susan Wright, "Recombinant DNA Technology and Its Social Transformation, 1972–1982," *Osiris* 2 (1986), 303–60; Jack Ralph Kloppenburg, *First the Seed: The Political Economy of Plant Biotechnology, 1492–2000* (Cambridge: Cambridge University Press, 1988); and Sheldon Krimsky, *Biotechnics and Society: The Rise of Industrial Genetics* (New York: Praeger, 1991). For a non-normative approach, contrast Joan H. Fujimura, "The Molecular Biological Bandwagon in Cancer Research: Where Social Worlds Meet," *Social Problems* 35 (1988), 261–83.

3. See, e.g., Roger Ekins, "A Shadow over Immunoassay," *Nature* 340 (1989), 256–58. For a recent journalistic example see Lawrence M. Fisher, "Profits and Ethics Clash in Research on Genetic Coding," *New York Times*, 30 Jan. 1994, pp. 1, 18.

4. Charles Weiner, "Universities, Professors, and Patents: A Continuing Controversy," *Technology Review* 86, 2 (Feb. 1986), 32–43.

5. Gerald E. Markle and Stanley S. Robin, "Biotechnology and the Social Reconstruction of Molecular Biology," *Science, Technology, and Human Values* 10 (1985), 70–79.

6. Sandra Panem, *The Interferon Crusade* (Washington, D.C.: Brookings Institution, 1984).

7. Michael Lynch, "Technical Work and Critical Inquiry: Investigations in a Scientific Laboratory," *Social Studies of Science* 12 (1982), 499–533. For a critique of the "sociology of accusation," and the shift from a "critical" sociology to a sociology of the actors' critical work, see Luc Boltanski, *L'amour et la justice comme compétences* (Paris: Métailié, 1990), esp. part 1, "Ce dont les gens sont capables." For a constructivist analysis of public controversies, see Alberto Cambrosio and Camille Limoges, "Controversies as Governing Processes in Technology Assessment," *Technology Analysis and Strategic Management* 3 (1991), 377–96.

8. Michel Callon, "Is Science a Public Good?" *Science, Technology, and Human Values* 19 (1994), 395–424.

9. See, e.g., Bruno Latour, *Science in Action: How to Follow Scientists and Engineers Through Society* (Cambridge, Mass.: Harvard University Press, 1987).

10. Joseph O'Connell, "Metrology: The Creation of Universality by the Circulation of Particulars," *Social Studies of Science* 23 (1993), 129–73. For a biomedical example, see Peter Keating and Alberto Cambrosio, "Interlaboratory Life: Regulating Flow Cytometry," in *The Invisible Industrialist: Manufacturers and the Construction of Scientific Knowledge*, ed. Jean-Paul Gaudillière, Ilana Löwy, and Dominique Pestre (London: Macmillan, 1996).

11. Very revealing, from this point of view, are the debates over which institution—private industry or public schools—should assume the burden of training highly specialized technicians; see, e.g., Elizabeth M. Useem, *Low-Tech Education in a High-Tech World: Corporations and Classrooms in the New Information Society* (New York: Free Press, 1986).

12. Richard A. Goldsby, S. Srikumaran, and Albert J. Guidry, "Cell Culture and the Origins of Hybridoma Technology," in *Hybridoma Technology in Agricultural and Veterinary Research*, ed. Norman J. Stern and H. Ray Gamble (Totowa, N.J.: Rowman and Allanheld, 1984), pp. 8–14, quoting from p. 12. The original paper was Georges Köhler and César Milstein, "Continuous Cultures of Fused Cells Secreting Antibody of Predefined Specificity," *Nature* 256 (1975), 495–97. For a detailed examination of the development of HT, see Alberto Cambrosio and Peter Keating, *Exquisite Specificity: The Monoclonal Antibody Revolution* (New York: Oxford University Press, 1995).

13. Grant Fjermedal, *Magic Bullets* (New York: Macmillan, 1984).

14. David Dickson, "Wistar Denied Monoclonal Antibody Patent in U.K.," *Science* 222 (1983), 1309.

15. "Supreme Court Upholds Hybritech Patent," *Biotechnology News*, 24 April 1987, p. 1; and "Hybritech Sues Abbott," *McGraw-Hill's Biotechnology News*, 15 Dec. 1986, p. 3.

16. "Becton, Ortho in Historic $5 Million Monoclonal and Cytometry Settlement," *Bioengineering News*, 10 Oct. 1986, p. 1.

17. Ekins, "Shadow" (note 3), p. 258.

18. Michael Mackenzie, Peter Keating, and Alberto Cambrosio, "Patents and Free Scientific Information in Biotechnology: Making Monoclonal Antibodies Proprietary," *Science, Technology, and Human Values* 15 (1990), 65–83.

19. Köhler and Milstein, "Continuous Cultures" (note 12). When the term "hybridoma technology" finally emerged, some observers objected to its eye-catching connotation. A would-be precursor of HT noted: "We did not propose in print that the fused cells may serve practical purposes as sources of antibodies; we did not call the fused cells 'hybridomas'." Joseph G. Sinkovics, "Early History of Specific Antibody-producing Lymphocyte Hybridomas," *Cancer Research* 41 (1981), 1246–47.

20. Alberto Cambrosio and Peter Keating, "Between Fact and Technique: The Beginnings of Hybridoma Technology," *Journal of the History of Biology* 225 (1992), 175–230. For the notion of "evidential context" see Trevor Pinch, *Confronting Nature: The Sociology of Solar-Neutrino Detection* (Dordrecht: Reidel, 1986).

21. Dale E. Yelton and Matthew D. Scharff, "Monoclonal Antibodies: A Powerful New Tool in Biology and Medicine," *Annual Review of Biochemistry* 50 (1981), 657–80.

22. César Milstein, "From Antibody Structure to Immunological Diversification of Immune Response," *Science* 231 (1986), 1261–68.

23. Interview of Melvin Cohn, 15 March 1990 (Salk Institute, La Jolla, California), by Alberto Cambrosio; and Timothy A. Springer, ed., *Hybridoma Technology in the Biosciences and Medicine* (New York: Plenum Press, 1985), p. xi.

24. For the notion of "experimental system" see Hans-Jörg Rheinberger, "Experiment, Difference, and Writing: I. Tracing Protein Synthesis," *Studies in History and Philosophy of Science* 23 (1992), 305–31; and Rheinberger, "Experiment, Difference, and Writing: II. The Laboratory Production of Transfer RNA," ibid., pp. 389–422.

25. Norman R. Klinman, "Antibody with Homogeneous Antigen Binding Produced by Splenic Foci in Organ Culture," *Immunochemistry* 6 (1969), 757–59. Some scientists disputed whether antibodies produced with Klinman's technique were proved monoclonal. The technique remained in competition with HT for a number of years. One text preferred Klinman's technique "if the intent is to examine the mouse repertoire of evokable antibody responses to human cell surface antigens, and to quickly define as many antigens as possible," but HT "if large quantities or stable sources of standard reagents are to be produced." See R. Levy, J. Dilley, and L. A. Lampson, "Human Normal and Leukemia Cell Surface Antigens: Mouse Monoclonal Antibodies as Probes," in *Lymphocyte Hybridomas*, ed. Fritz Melchers, Michael Potter and Noel L. Warner, "Current Topics in Microbiology and Immunology" 81 (Berlin: Springer Verlag, 1978), pp. 164–69.

26. Interview of Walter Gerhard (Wistar Institute, Philadelphia), 28 Sept. 1987, by Alberto Cambrosio and Peter Keating.

27. Interview of William Wunner (Wistar Institute, Philadelphia), 30 Sept. 1987, by Alberto Cambrosio and Peter Keating.

28. Roger P. Ekins, personal communications, 27, 31 Jan. 1994.

29. Interview of Leonard A. Herzenberg and Lee A. Herzenberg (Stanford, California), 27 March 1988, by Alberto Cambrosio, here referring to Leonard A. Herzenberg and Jeffrey A. Ledbetter, "Monoclonal Antibodies and the Fluorescence-Activated Cell Sorter: Complementary Tools in Lymphoid Cell Biology," in *The Molecular Basis of Immune Cell Function*, ed. J. Gordin Kaplan (Amsterdam: Elsevier/North Holland Biomedical Press, 1979), pp. 315–30.

30. On the notion of opportunities, see Joseph Rouse, *Knowledge and Power: Toward a Political Philosophy of Science* (Ithaca, N.Y.: Cornell University Press, 1987), pp. 80–95.

31. Alberto J. L. Macario and Everly Conway de Macario, "Introduction: Monoclonal Antibodies against Bacteria for Medicine, Dentistry, Veterinary Sciences, Biotechnology, and Industry—An Overview," in *Monoclonal Antibodies Against Bacteria*, ed. Macario and Conway de Macario, 3 vols. (Orlando, Fla.: Academic Press, 1985), vol. 1, pp. xvii–xxxiii.

32. National Institute of Allergy and Infectious Diseases (NIAID), Immunology Study Group, *New Initiatives in Immunology* (NIH Publication No. 81-2215) (Bethesda, Md.: U.S. Department of Health and Human Services, 1981).

33. B. Kahan, R. Auerbach, B. J. Alter, and F. H. Bach, "Histocompatibility and Isoenzyme Differences in Commercially Supplied 'BALB/c' Mice," *Science* 217 (1982), 379–81; H. L. Foster and M. W. Balk, "Histocompatibility and Isoenzyme Differences in Commercially Supplied 'BALB/c' Mice: A Reply," ibid., p. 381; and "Mongrel Mice Caused No Major Monoclonal Losses, Hybridoma Makers Report," *McGraw-Hill's Biotechnology Newswatch*, 16 Aug. 1982, p. 5.

34. On the artificial nature of laboratory materials see Karin D. Knorr-Cetina, *The Manufacture of Knowledge: An Essay on the Constructivist and Contextual Nature of Science* (New York: Pergamon Press, 1981); Robert E. Kohler, *Lords of the Fly: Drosophila Genetics and the Experimental Life* (Chicago: University of Chicago Press, 1994); and Nelly Oudshoorn, "On the Making of Sex Hormones: Research Materials and the Production of Knowledge," *Social Studies of Science* 20 (1990), 5–33.

35. NIAID, *New Initiatives* (note 32).

36. Alberto Cambrosio and Peter Keating, "A Matter of FACS: Constituting Novel Entities in Immunology," *Medical Anthropology Quarterly* 6 (1992), 362–84.

37. Alain Bussard, Micah I. Krichevsky, and Lois D. Blaine, "An International Hybridoma Data Bank: Aims, Structure, and Function," in Macario and Conway de Macario, *Monoclonal Antibodies* (note 31), pp. 187–311.

38. Herzenberg and Herzenberg interview (note 29).

39. "The established cell lines offer the further advantage of unlimited permanent supply of material, and the possibility of worldwide standardization," Giovanni Galfrè, S. C. Howe, César Milstein, G. W. Butcher, and J. C. Howard, "Antibodies to Major Histocompatibility Antigens Produced by Hybrid Cell Lines," *Nature* 266 (1977), 550–52.

40. Dale E. Yelton, David H. Margulies, Betty A. Diamond, and Matthew D. Scharff, "Plasmacytomas and Hybridomas: Development and Applications," in *Monoclonal Antibodies: Hybridomas, a New Dimension in Biological Analyses*, ed. Roger H. Kennet, Thomas J. McKearns, and Kathleen B. Bechtol (New York: Plenum Press, 1980), pp. 3–17.

41. Melvin Cohn, 1981 National Cancer Institute contract proposal. Copy courtesy of Melvin Cohn.

42. Herzenberg and Herzenberg interview (note 29).

43. Interviews of Patrick Kung (T-Cell Sciences, Cambridge, Mass.) 15 July 1993; and of Stuart F. Schlossman (Dana Farber Cancer Institute, Boston), 30 April 1992; by Alberto Cambrosio and Peter Keating.

44. Herzenberg and Herzenberg interview (note 29).

45. Henry Krakauer, "Central Banking of Hybridomas," *In Vitro* 17 (1981), 1081–83.

46. Ibid., p. 1083. For the patent issues see, e.g., H. Boy Woodruff and Brinton M. Miller, "Patenting of Hybridomas and Genetically Engineered Microorganisms," *In Vitro* 17 (1981), 1078–80; Dale H. Hoscheit, "United States Patent Requirements," ibid., pp. 1084–85; and Irving Marcus, "International Patent Procedures," ibid., pp. 1086–88.

47. Gaston Bachelard, *Le matérialisme rationnel* (Paris: Presses Universitaires de France, 1953), pp. 71–81.

The Human Genome Project and the Acceleration of Biotechnology

Michael Fortun

There is something almost palpable about the rhetoric of velocities permeating and surrounding the enterprise known as the Human Genome Project. My aim in this essay is to provide some limited, concrete examples of how speed was talked about during the construction of the Human Genome Project (HGP): what it was, how it could be achieved, why it was so incredibly desirable. The intent is to begin an ethnography of speed in the field of genomics, to initiate a process of inquiry into the "lost dimension" of speed.[1] For it has by and large escaped our theorizing, despite its insinuation into nearly every nook and cranny of genomics discourse from the laboratory to the congressional hearing room. Why, for example, is Big Science a fairly stable category of thought in science studies, making for a great deal of debate about whether the Human Genome Project is or is not Big Science, while Fast Science does not appear as a matter of public or even disciplinary concern?[2] Speed remains incidental to the "real" issues of knowledge production and the "ethical, legal, and social issues" of genomics, an accidental quality that need not be discussed or worried over.

As one consequence, speed often appears a matter of choice while the development of genomics itself retains a certain quality of inevitability. For example, in 1989 Robert G. Martin, chief of the microbial genetics section in the Laboratory of Molecular Biology of the National Institute of Diabetes and Digestive and Kidney Diseases, raised many issues in an article critical of the HGP, but speed stood out. "Why the race to complete the mapping of the human genome in 5 years? . . . I do not deny that mapping the human genome is worthwhile. I do question whether there is justification for the urgency it is receiving. . . . I simply cannot fathom the haste." Or, as the Harvard microbiologist Bernard Davis asked succinctly in the same issue of *NIHAA Update*: "What's the Big Hurry?"[3] So while I may agree with the spirit of these comments (or with Evelyn Fox Keller when she states that "the real controversy about the Human Genome Project is about pace, it's about the speed. Do we spend $3 billion on it or do we just let it happen a little more slowly?"[4]) I also want to question whether (and how) speed in this case can be a matter of public deliberation. I will argue that speed is not incidental to the HGP but is constitutive of it; a Human Genome Project done "a little more slowly" would not be the Human Genome Project. The common journalistic description of the HGP as "a project to map and sequence the entire human genome" is not only inaccurate but incomplete; it should at least be recast to read "a project to map and sequence the entire human genome *as fast as we possibly can*."[5]

How does speed, then, distinguish the Human Genome Project from what came before it, or what might have replaced it? What produces speed, and what limits it? Where are its sources? In what situations is speed "of the essence"? In what situations is it sacrificed for other considerations? Most difficult of all, what are its political effects and meanings?

Making Speed in the Laboratory

By beginning in the laboratory I do not mean to suggest that the origins of speed (still undefined) are to be found there—that is, that the technical practices of genomics are the primary source, the first link in a causal chain of speed. The rhizomatic network of genomics has no such center, but is best thought of as an example of what Gilles Deleuze and Felix Guattari call "acentered systems, finite networks of automata in which communication runs from any neighbor to any other, the stems or channels do not preexist, and all individuals are interchangeable, defined only by their *state* at a given moment—such that the local operations are coordinated and the final, global result synchronized without a central agency."[6]

But the laboratory is as good a place as any to begin tracing out the intricacies of the genomics speed constellation, especially since it has been the productive focus of some recent work in the sociology of science. Joan Fujimura has elaborated how the oncogene "bandwagon" was constructed with the aid of the standardized packages of theory, instruments, and craft skills of molecular biology.[7] Implicit in the vehicular metaphor of *bandwagon* is a notion of speed: under certain conditions—that is to say, within a given acentered assemblage of those standardized packages, plus National Institutes of Health funding patterns, career demands, constraints of law and capital, and other sociotechnical elements articulated into quasi-stable working patterns—oncogene research can develop more speed and momentum than, say, research on environmental carcinogens. This is not to say that such tools always work the way they are supposed to, or do not require extensive coaching, fiddling, personal skill, or luck. Examining how the technique of "plasmid preps" (plasmid purification and isolation) is practically accomplished, Kathleen Jordan and Michael Lynch analyze just how slow these fast technologies can sometimes be.[8] But for the present purpose of exploring high-speed configurations, I leave such considerations aside.

Throughout the history of genetics, elements of speed have been evident in investigators' choices of experimental organisms: the relatively rapid reproductive cycles of *Drosophila*, *Neurospora*, *Escherichia coli*, and bacteriophage have figured significantly in the laboratory success of these organisms and their attendant human actors. Humans historically made for poor genetic research, according to modern genomicists David Botstein and Eric Lander, for two reasons: "(1) Because *Homo sapiens* is not an experimental animal that can be manipulated at will and (2) because few genetic markers had been found that were heterozygous *often enough* to allow random matings to be informative."[9] The technology that allowed one to work around this speed problem implicit in population genetics involved the creation and use of restriction fragment length polymorphisms (RFLPs), a methodology worked out by Botstein and his colleagues in the late 1970s.[10] While early reports of their research in this area were inconclusive about the technique's ultimate feasibility for human genetics, the practical implications were already being discussed, as at a Banbury symposium at Cold Spring Harbor in 1979. A series of exchanges between Ray White and John Cairns published in the proceedings from that conference provide some clues as to what this new technology would mean for the practice of human genetics, and how this practice was beginning to be considered in its economic context—and as an activity with its own economics, particularly an economics of time:

Cairns: I would like to interject again that I think there's a real difference that the classical epidemiologists and epidemiological geneticists should think about. When they contemplate action on collecting some specific data, they consider the cost and whether or not it is practical. In this molecular business, the thought is farther from the action, and there's a whole different approach.
White: So you're saying the response time in this business is relatively short.
Cairns: Yes.
White: I agree.
Cairns: In principle, if you can think of it you can do it.
White: Yes. The lag time between the selection scheme and trying it, for example, is about a week.[11]

RFLP is just one technology in a genomic armamentarium that has grown to include the polymerase chain reaction (PCR), fluorescent in situ hybridization (FISH), souped-up vector vehicles such as yeast artificial chromosomes (YACs), and dozens of other cyborg assemblages, including the centerpiece of speed obsession and production, the DNA sequencer. Such standardized tools and biomaterials for genomics research are now advertised on the basis of their ability to create speed, to reduce the time quotient in the equations of efficiency (see Figures 1, 2). A new market niche of what

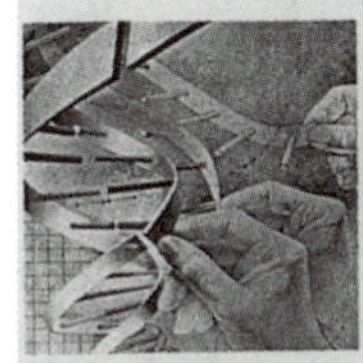

ONE-HOUR DNA

NEW UltraFAST DNA Synthesis Technology

Why wait hours or days for your oligonucleotides when you can have them in less than one hour? Beckman now offers you a way to remove the oligo bottleneck and increase your laboratory efficiency with Ultra*FAST* DNA Synthesis.

FAST, EASY DNA SYNTHESIS

All you need for Ultra*FAST* DNA Synthesis is our Oligo 1000 DNA Synthesizer running new *FAST* synthesis methods (2-minute cycle time for a 30 nmole scale and 2.25 minutes for a 200 nmole scale) and our unique acetyl-protected deoxyC phosphoramidite*. Then use our new Ultra*FAST* Cleavage & Deprotection Kit* which completely processes your DNA in ten minutes. Take another ten minutes to dry a useful aliquot and you have ready-to-use DNA in less than one hour.

BECKMAN MAKES IT SIMPLE

With our *FAST* DNA synthesis methods for two-minute cycle times and new DNA chemistries, the Oligo 1000 DNA Synthesizer is the instrument of choice for fast, easy and economical DNA synthesis. *Dr. Len Packman, Assistant Director of Research, Cambridge Centre for Molecular Recognition, Cambridge University, U.K. comments,* ***"The Oligo 1000 has consistently produced high quality oligonucleotides and has impressed me with both its ease of use and its speed of synthesis. It would be a valuable addition to any laboratory needing a fast and regular supply of primers and probes."***

THE BECKMAN PLUS

Beckman provides the science, support and solutions to help you in your important gene research. Call your local Beckman representative to get *free* technical information on how you can increase your laboratory efficiency with the new Ultra*FAST* DNA synthesis technology from Beckman.

*Patents Pending.

DNA in less than an hour!
Think of the possibilities.

BECKMAN

Worldwide Offices: Africa, Middle East, Eastern Europe (Switzerland) (22) 994 07 07. **Australia** (61) 02816-5288. **Austria** (2243) 85656-0. **Canada** (905) 387-6799. **China** (861) 5051241-2. **France** (33) 1 43 01 70 00. **Germany** (49) 89-38871. **Hong Kong** (852) 814 7431. **Italy** (39) 2-953921. **Japan** 3-3221-5831. **Mexico** 525 575 5200, 525 575 3511. **Netherlands** 02979-85651. **Poland** 408822, 408833. **Singapore** (65) 339 3633. **South Africa** (27) 11-805-2014/5. **Spain** (1) 358-0051. **Sweden** (8) 98-5320. **Switzerland** (22) 994 07 07. **Taiwan** (886) 02 378-3456. **U.K.** (0494) 441181. **U.S.A.** 1-800-742-2345.

© 1993 Beckman Instruments, Inc.

Circle No. 20 on Readers' Service Card

Figure 1. Speed equals efficiency. Beckman Instruments "UltraFAST DNA synthesis technology" is "fast, easy, and economical."

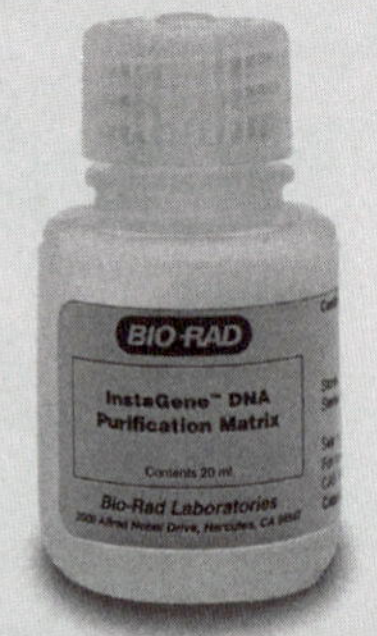

Figure 2. Bio-Rad Laboratories InstaGene matrix "not only speeds up isolation, but also helps you cut costs."

the trade journal *Genetic Engineering News* called "downright boring" products and processes has been created as a service sector to the genomics project. Companies like Bios Laboratories (which claims to be the "top supplier of experimental biologicals to HGP researchers") build "better mousetraps for the basic steps in molecular genetics," according to its chief operating officer. "This is speeding up the research clock. Instead of taking a couple of years, researchers are taking a month."[12]

But the speed with which these technologies of speed were themselves changing was (and remains) cause for a whole set of organizational concerns among genomics project managers. Paul Virilio's general observation that in modern speed regimes the faster engine "is often obsolete even before being used; the product is literally worn out before being operated"[13] is certainly applicable to genomics. Genomicist-administrators such as Charles Cantor are acutely aware of their precarious and ephemeral position between obsolescence and nonexistence:

> Just in the first two years . . . the technology is changing dramatically. It's changing so quickly that original attempts to sort of computerize and automate some of the first methods have become obsolete, because the methods changed before the program could be written. One of the areas that has changed dramatically is the methodology used for mapping. [You will learn of] direct or indirect progress towards a number of new, totally different types of methods, including doing mapping directly by sequencing, [radiation hybrids, and single-sperm PCR]. . . . This really is a partial list of methods that a couple years ago we didn't really anticipate, and they are already changing tremendously the way that this project's going to proceed.[14]

The organizational problems associated with speed did nothing to quench the genomic speed thirst and indeed, as I shall argue shortly, seem only to have heightened the desire. C. Thomas Caskey's remarks at a 1987 Office of Technology Assessment workshop on "issues of collaboration" are saturated with cravings for speed in the laboratory. After alluding to the "substantial errors" made by recombinant DNA researchers in "having developed this very powerful technology in the United States" and then "immediately put the brakes on" in the 1970s, Caskey made it clear that he wanted to avoid any further sluggish impulses:

> Every molecular biologist in the United States has known "clone by phone" for years. The communication and exchange between investigators in the United States has kept the research at an extremely high tempo. . . . Professor X at Stanford has something that you've heard about at a Gordon Conference or a meeting and it would greatly facilitate your research in an area that is probably not identical to his own interest. You ring up the professor at Stanford, the next day in the mail by Federal Express comes the clone and you're off and running. You've cut back nine months, twelve months of developmental work and you're immediately moving on your project. . . .
>
> . . . I am definitely detecting, as I call my friends now and discuss these things, a tightening of this attitude. I would like to see us focus on ways of facilitating the exchange of information materials. And if we have cumbersome old patent laws that

aren't staying up to speed with the speed of this technology and science, then let's rewrite it all, let's get it straight, let's maintain the leadership, let's facilitate these actions.[15]

This brief excerpt points to many other nodes of the genomics speed assemblage that require attention for a thorough understanding of what speed is, how it is achieved, and what it means—links to telecommunications technologies; dapper uniformed Federal Express agents transporting styrofoam containers of carefully prepared biomaterials between what we still quaintly distinguish as "private corporate labs" and "academic" or "government" laboratories; even more dapper uniformed attorneys transporting bits of text to create new ways to streamline "cumbersome old patent laws" or regulatory policies;[16] the relentless subdivision of research specialties into problems and trajectories "probably not identical" to those of the ever-present competitors; and that most deliciously vague allusion to "tightening of attitudes" regarding the sharing of biomaterials and bioinformation. Some of these topics will be touched on later, but for now they would slow this text down too much.

Problems of Speed as an Origin of the Genomics Project

It was in part the success of the new genomic technologies of the early and mid-1980s in producing speed that made something like the Human Genome Project seem necessary. One indicator of this relationship to speed can be found in the Human Gene Mapping (HGM) workshops, begun in 1973 as a way of compiling, organizing, and sharing results and information on the gene and genetic maps being built from the increasing data generated by new cytogenetic and molecular technologies. These roughly biannual workshops underwent rapid and dramatic change with the introduction of RFLP and other DNA-based technologies around 1980. Committees for specific chromosomes began proliferating, while new committees on nomenclature and DNA methodologies were created. The number of genes and cloned arbitrary DNA segments being mapped increased by as much as sixfold between 1981 and 1983, and the number of conference participants doubled.[17] By the 1985 meeting in Helsinki, the total number of sites mapped (1,479) had nearly doubled again, as had the number of cloned genes. The most immediate and pressing need, however, was for a "system for permanent storage and updating of human gene mapping data." Each previous workshop had "acted on its own, thus unnecessarily duplicating earlier work, especially with regard to typing and tabulating."[18] Informatics was becoming the most important speed nexus. The workshop's Committee on Human Gene Mapping by Recombinant DNA Techniques announced that "a major restructuring of this committee's methods of data collection and analysis will be required so that compilation can be continuous between now and the next workshop." Data handling could no longer be done on "an *ad hoc* basis. Extensive databases will need to be created, maintained continuously, and reviewed at the workshops."[19]

Computers had first been used at the Los Angeles workshop two years before, but they were used primarily as word processors. Not until HGM 9 in Paris in 1987 was the first "true database" used, and even then it could only be searched, not modified on-site. Only when the workshop arrived back at Yale for HGM 10 in 1989 was an interactive, on-line database in use. By that time the workshops were in "transition," according to Frank Ruddle and Kenneth Kidd, and the growing importance of computers best exemplified the nature of these changes in "a critical period." Now that the U.S. government had announced a commitment to an organized mapping and sequencing project, "the single most problematic aspect" of such a project was to define and develop "new organizational structures to enable the international community to integrate information into the ever-changing map in a timely, accurate, and accessible form." "The task of monitoring and reporting construction of the human gene map," Ruddle and Kidd suggested, "has become a full-time job."[20]

And, in an early expression of the add-fuel-to-the-fire philosophy that would be fully realized with the creation of the Human Genome Project, workshop organizers discussed how to accelerate further this already rapidly advancing specialty. They argued for increased sharing and standardization of materials—libraries, somatic cell hybrid panels, lymphoblast cultures—as one means of speeding up work results. Even cloned probes, they suggested—the fundamental unit of mapping work and the basis for securing scientific credit and patentable products—should be exchanged, at least after a laboratory's "particular project" had been completed.[21] The solution to the problems of speed was to go faster.

The problems of speed and informatics were expressed in some of the first legislation that sought to institutionalize the HGP in the United States, the Biotechnology Competitiveness Act of 1987. The centerpiece of this bill was the establishment of the National Center for Biotechnology Information at the National Library of Medicine. The swell and rush of new sequence data being generated with new standardized molecular genetic packages had by 1984 resulted in a twelve-month delay between data publication and entry into GenBank, the Department of Energy's database, then in operation for two years. GenBank administrators hired additional clerical workers and invented new arrangements with authors and journals. By 1986 the time delay was down to six months, but further data acceleration was already foreseen with the then-hoped-for Human Genome Project.[22] As the 1987 bill stated: "Information on the map and sequence of the human genome, as well as the many other important fields of biotechnology research, is accumulating faster than can be reasonably assimilated by present methods," making it "essential that advances in information science and technology be made so that this vast new knowledge can be organized, stored, and used" — to organize, store, and use still more data, faster still.[23]

Defining the Human Genome Project in Terms of Speed

The critiques of speed that Martin and Davis published in 1989[24] appeared when the Human Genome Project was under heightened attack, albeit the criticism was moot, since the National Center for Human Genome Research (NCHGR) had just recently been established within the NIH, with James Watson as its first director.[25] Davis was nevertheless invited to testify before a U.S. Senate committee in 1990 to air his criticisms of the HGP as redefined and renegotiated over the previous few years by a relatively small group of genomics proponents meeting under changing institutional labels, among them the National Academy of Sciences panel and the NIH Ad Hoc Program Advisory Committee on Complex Genomes. While Davis lauded the de-emphasis of sequencing and the new stress on mapping, informatics, and the analysis of model organisms, he questioned whether this was anything so innovative as to merit special treatment as a separate NIH granting mechanism:

> These are all excellent goals. But except for the mapping I would suggest that they are what we would be doing today if there were no human genome project. We would be doing all of these things through competing grants, though no doubt on a smaller scale. If this project were to be submitted today as something new, without the history of the original political appeal of the very definite goal of something like putting a man on the moon — getting that last human nucleotide — if it were to be proposed today as a new project, it seems doubtful that it would generate the same political appeal that the original one had.[26]

Davis alludes here to a persistent definitional problem that genomics proponents had faced. What was the difference between a "Human Genome Project" and business as usual, that is, competing grants "on a smaller scale"? As we have seen, that "smaller scale" was already producing more data than could be managed effectively. Arguments made for the Human Genome Project in the mid-1980s boiled down to speeding up what was already going on.

This fact was evident in Walter Gilbert's introductory remarks at a Cold Spring Harbor session in 1986 — the first at which the larger community of molecular biologists discussed the desire to map and sequence "the" human genome completely. Gilbert laid out three possible time horizons: centuries at the present rate of sequencing; a "moderate period of time," roughly twenty or thirty years, if one "focus[ed] some level of effort at the problem"; and finally:

> The third possibility is to try to do this in, let's say a scientist's immediate lifetime, a much shorter time than this, and let's say, let us try to develop a map of the human

genome, and a sequence, by the year 2000. Fifteen years from now. I think the third is a—the third is a more striking proposal. The third is a proposal in which if you discuss what you could learn from looking at the human genome, it's something which you can imagine being here in a reasonable time.

"I doubt one can energize around" the first proposal, Gilbert reflected. Like many "moderate" proposals, his second option faded quickly from consciousness as well. Gilbert and a few others would spend the next several years trying to convince the appropriate people, trying to "energize around" the idea that "a scientist's immediate lifetime" was indeed a "reasonable time."[27]

A number of arguments were advanced in support of this high-speed option. Occasionally they were made in the most abstract and general terms, as in the remarks G. Kirk Raab, President and CEO of Genentech, made at a congressional hearing in support of the HGP:

> At Genentech . . . we support 100 percent of this effort, and I briefly want to tell you why.
>
> First, speed. The speed which this will give us to solve problems can be likened to what has happened in the computer sciences, and what we can do today, as compared to just five and ten years ago, the speed of science is tremendous, and the ability to use technicians rather than scientists to make things happen.[28]

Granted, government hearings are not always the occasion for specificities and direct remarks; still, speed seems to have escaped specific rational justification—or rather, to evoke something beyond those rational justifications offered. That speed can be *built in*, so that the skills of the scientist can be replaced by those of the technician at a reduced labor cost: one can readily understand the president of Genentech making such a calculative rationale. But something at once simpler and more complex seems to be expressed here, an apparent desire simply *to go fast*.

There is little doubt that the requirements of capital investment play an important part in the genomics speed assemblage and arguments for it, a point made again below.[29] But by far the most frequent and effective speed argument involved the potential medical importance of genomics-based knowledge and practices. In late 1985, when the scientific and bureaucratic appetite for a large-scale, organized genomics project was still to be found only in the U.S. Department of Energy, Charles DeLisi e-mailed a colleague a few thoughts that presaged the comments of Gilbert cited above. DeLisi replied to one of the anticipated criticisms that his colleague had voiced, "With respect to the grind, grind, grind . . . argument, it seems to me that if the human genome is to be sequenced, there will be some grind work; what we are discussing is whether the grinding should be spread out over thirty years (say), or compressed into ten. I think a case can be made that the impact on medicine and biology will be so enormous relative to the costs of this project that acceleration is overwhelmingly desirable."[30]

Such arguments also goaded participants to keep a rapid pace once the project was under way, and meshed well with other speed incentives. "We have to get some real results in the next five years . . . find the gene for something which you might not have found if you didn't have the human genome mapped," James Watson said in regard to the need to keep congressional patrons interested and enthusiastic. But he quickly joined this political imperative to an ethical one: "If you want to understand Alzheimer's disease, then I'd say you better sequence chromosome 21 as fast as possible. And it's unethical and irresponsible *not* to do it as fast as possible."[31] Or, as he put it to a Senate committee,

> Some people say, well, why this targeted project, why the hurry? Why not just let the individual investigator (RO1) grants come in, and not have this big program. I think if you are a family with a serious genetic disease, you know, there are only a certain number of years you are going to live. You do not want that gene to be found 100 years after you are dead. So, if we can do this work over the next five years, I think we should do it. . . . I think we are doing the moral thing to try hard to get these genetic resources . . . if you can find it now, I would rather have it this year instead of ten years from now. So I am in a hurry.[32]

To this linking of medical and moral imperatives, we should add personal motivations. I have mentioned Gilbert's urging his colleagues at Cold Spring Harbor in 1986 to do the project "in a scientist's immediate lifetime," that is, the lifetime of a scientist of his particular generation. Watson too has mused upon his own mortality, as well as that of members of families with serious genetic diseases, and invoked the narrative demands of his life and career:

> People ask why *I* want to get the human genome. Some suggest that the reason is that it would be a wonderful end to my career. . . . That *is* a good story. . . . The younger scientists can work on their grants until they are bored and still get the genome before they die. But to me it is crucial that we get the human genome now rather than twenty years from now, because I might be dead then and I don't want to miss out on learning how life works.[33]

The Speed of Politics and the Politics of Speed

> To govern is more than ever to fore-see, in other words, to go faster, *to see before.*
>
> —Paul Virilio[34]

As Barney Glaser and Anselm Strauss have pointed out in their exemplary research on the temporal aspects of "status passages," "*The legitimator who must create a timing for a seldom tried, new, or temporally unknown passage has more discretion in setting the timing* than has a legitimator for a passage wherein the timing is generally well known, and typically scheduled and organized—e.g.

a school year."[35] As head of the NCHGR and thus chief legitimator for the Human Genome Project, Watson—whose own career was founded on the superior velocity exhibited in the "race" for the double helix—helped speed up genomics by means of his advocacy for this enlarged, accelerated, and independent funding mechanism with the fancy name. Just as Gilbert's introductory intervention at the 1986 Cold Spring Harbor meeting pushed the "fast track" option, Watson worked the molecular genetics community, the halls of the U.S. Congress, and the official committees he helped establish and on which he served, to keep genomics moving apace. Of the many instances in which Watson flexed his political muscles, I will mention just one episode. Recall that at the Cold Spring Harbor meeting Gilbert set out three possible futures that varied in speed and commitment of resources; in that situation, the fastest and largest proposal was made out to be the one that "energy could be organized around." Robert Cook-Deegan describes a similar scenario, played out at a meeting of the National Academy of Sciences/National Research Council (NAS/NRC) committee; it too involved the level of funding and, tied to that, the speed at which events would proceed:

> A subgroup was tasked to produce cost options. Botstein spearheaded this effort and came up with three options that would enable various amounts of genetic linkage mapping, physical mapping, sequencing, and other activities. He suggested annual budgets of $50 million, $100 million, and $200 million, with completion dates sooner for the higher figures (the year 2000 for $200 million versus 2025 for $50 million). . . . Watson objected to the range of options, noting that it would naturally incline the committee members to seem reasonable by choosing the middle option. He therefore suggested a $500 million-per-year crash program as a fourth option. (Because Botstein had already deemed his top option the crash program, Watson's was variously dubbed the crash-crash or crash-boom.)[36]

Watson appears as a savvy politician, well-versed in the dynamics of group decision making, intervening at the appropriate time to keep things moving in certain directions. After some discussion among the subgroup and a "review" of figures by Botstein, the committee decided to recommend $200 million per year.

Standardized genomics packages do not alone produce speed, but require in addition this kind and level of capital commitment.[37] This externalized "crash-boom" speed of funding level was also *internalized*, and demarcated the Human Genome Project from what Davis in 1990 argued would be done anyway. On the NAS/NRC panel, as Cook-Deegan has reported, the yeast genomicist Maynard Olson argued that "projects should be considered genome research only if they promised to increase scale factors by threefold to tenfold (size of DNA to be handled or mapped, degree of map resolution, speed, cost, accuracy or other factors)."[38] Acceleration thus defined HGP research. The NAS/NRC committee adopted Olson's definition in its 1988 report, so important to establishing an image of community consensus on the Human Genome Project: "The human genome project should differ from present ongoing research inasmuch as the component subprojects should have the potential to improve by 5- to 10-fold increments the scale or efficiency of mapping, sequencing, analyzing, or interpreting the information in the human genome."[39]

Who stood in the best position to benefit from speed? What kinds of structural advantages might exist, emerge, or be created by defining "the Human Genome Project" in terms of speed? That definition led to a number of structural inequalities, or so some charged. Genomics proponents of a certain type and at a certain level of lab size and achievement—Maynard Olson, Leroy Hood, Walter Gilbert, C. Thomas Caskey—had lobbied the project into existence; they were also those able to provide the new speed levels sought; and the project was in fact defined largely in terms of what they were capable of and interested in doing. Critics charged that too much money was going to large center grants. After the HGP was established, Watson wanted an even "faster and more accountable genome project" and advocated an even more pronounced emphasis on contracts and centers: "It's my belief that the program has got to change emphasis. To get the job done, we need contracts, not grants."[40] To what extent did speed entail a certain level of cronyism, the strengthening and further stratification of speed hierarchies?

The question is difficult to answer. In the course of my fieldwork on the Human Genome Project, on several occasions I heard stories about so-and-so from such-and-such institution getting a grant on a proposal that had received a low priority score but was funded because of connections with someone on the advisory panels or study groups. Because such remarks are generally made off the record, they do not get built into our knowledge about the political culture of genomics. Earl Lane, a reporter on the genomics beat for the New York daily *Newsday*, also heard a number of such stories. Lane was researching how genomics proponents who served on important "advisory" and "evaluative" panels to the HGP "stand to receive millions of dollars over the planned fifteen year life of the program."[41] He requested through the Freedom of Information Act the raw priority scores of applicants to the NCHGR, a request that was denied; Lane appealed, and the appeal was denied by the Assistant Secretary for Health on the basis that, contrary to Lane's claims, such scores were evaluative in nature and their release would violate the privacy of researchers.[42]

Speed Limits

Limits were on occasion placed on speed. In extensive debates over ownership of information and materials and how this promoted international competitiveness, the "pure speed" position would have been to sequence and publish immediately; if the goal was to "get the job done" as quickly as possible, because of a moral imperative to cure disease or some other motive, the "logical" path to take was to make new information immediately available to everyone. The path actually taken has of course been more complicated, especially in respect to the management and sharing of data, as Stephen Hilgartner's paper in this volume shows. True, researchers at the genome center at the Whitehead Institute, for instance, honor such motivations by describing themselves as ready to make their mapping data "available prior to publication, as soon as it is confirmed," and not planning "to seek patents on the clones markers or maps, which they consider basic infrastructure for the scientific community." Whitehead researchers also generate speed by contracting out the bulk-synthesizing of PCR primer sequences for markers; they essentially give away half of these to other researchers: "instant access at affordable prices," as the center's director Eric Lander phrased it.[43]

Although sharing information and materials in this manner might increase not only the raw speed of research, but also the speed with which discoveries are commercialized, there are other commercial rationales that suggest the need for speed limits. At a congressional hearing in 1989, then-Senator Al Gore questioned Watson at length about the "practicality" of sharing information, particularly with Japan. Watson acknowledged that sharing information was "a very difficult issue." "But there are people," he continued,

> who would say that even before anyone begins to interpret the data, we should put it in a database so anyone in the world can do it. Because that will speed up science, and that will be good for the world. That is one argument. The other argument is that—or two others: it is commercially valuable, or, second, that the people who have done it want some scientific sense of achievement. And we are going to talk about this for many years to come.[44]

Speed, so highly and persistently valued, can be sacrificed in some situations. Making all information immediately available would speed up science and be "good for the world," but speed must sometimes take a back seat to protecting career credit and economic interests.

Speed in the service of those interests seems to have no limit. Genomicists now compete in what *Science* magazine has called "a mad scramble for scientific precedence," with both authors and publishers trying to outrace the competition. *Science* described one speed scenario as follows: Francis Collins (who had replaced Watson as director of the NCHGR) submitted his group's paper on neurofibromatosis to *Science*. Ray White, searching for three years for this same gene and now "furiously writing," phoned the editor of *Cell*, Benjamin Lewin, who accepted White's rapidly produced paper "a mere 17 days after submission." Learning that White's paper would thus appear first, Collins appears to have "alerted [*Science* editor Daniel Koshland] to the problem." Koshland then "went to the extraordinary measure of remaking the last few pages of the magazine to squeeze the Collins paper in a week ahead of schedule. . . . [B]oth groups shared the limelight at a well-attended press conference and on the front page of the *New York Times*."[45]

One notorious example of speed for commercial purposes is Craig Venter's scheme, developed while working at NIH, to pioneer the "business of spewing out sequences of gene pieces" known as cDNAs and then patenting those fragments of expressed genetic sequences as tools for further biomedical research.[46] It has received a lot of attention, most of it focused on such issues as whether cDNAS are "really useful" and hence "really" patentable, or on the deplorable fact that, according to some professional bioethicists (whose own importance develops in speedy tandem with genomics), "This is not science. This is like the Gold Rush"—as if that was not precisely the point. Venter's scheme epitomizes the genomics speed assemblage, combining standardized packages, fluid and abundant capital, and some degree of career building into a "land grab" of impressive proportions.[47]

"Venter's work highlights the speed of the human genome project," said the president of the Association of Biotechnology Companies, Forrest Anthony, and it was the unleashing of such pure speeds and the consequent unregulated competitive struggles that caused such great concern among the more moderate speed enthusiasts. Even the origin of this plan, journalistically reconstructed in the favored "Eureka!" format, is situated in persistent speed vectors. "Frustrated by the snail's pace at which he was able to decipher the human genetic material," Venter struggled for years to solve the speed problem when, "on an airplane coming home from a meeting in Tokyo," inspiration hit.[48]

But inspiration alone was insufficient and had to be joined to the high-speed, standardized, automated genomics packages discussed earlier. "Three commercial human brain cDNA libraries," marketed by Stratagene as lambda-ZAP catalog numbers 936206, 936205, and 935205, were "converted en masse to pBluescript plasmids" (a Stratagene vector system); "Qiagen columns (Studio City, California) improved the percentage of plasmid templates that yielded usable sequences;" these templates were used in the PCR process, and the results of that speedy reproduction were run through the Applied Biosystems DNA sequencer. The resulting sequences were examined for similarities in GenBank, and their possible protein analogues were matched against the Protein Information Resource database. Those searches were conducted "with our modifications of the 'basic local alignment search tool' programs for nucleotide (BLASTN) and peptide (BLASTX) comparisons"—programs based on an algorithm which "sacrific[es] some sensitivity for the 60- to 80-fold increase in speed over other database-searching programs such as FASTA."[49]

This was all too fast to be contained by the National Institutes of Health. With $70 million in venture capital that their speed technologies had at-

Figure 3. This advertisement for Genset products promises to compress the work of a billion years into less than a day.

tracted, Venter and thirty colleagues (nearly his entire staff) transferred their efforts to the new Institute for Genomic Research and its profit-making arm, Human Genome Sciences, Inc., at midnight, 13 July 1991. As the *New York Times* contextualized it:

> Wallace Steinberg, chairman of the board of HealthCare Investment Corp., which is financing Dr. Venter, said he suddenly realized that there was an international race to lock up the human genome. If Americans do not participate, he said they will forfeit the race and lose the rights to valuable genes to Britain, Japan and other countries that are in the race to win. He said the National Institutes of Health could not afford to invest enough money in Dr. Venter's enterprise to make it truly competitive.[50]

Steinberg's comments here echo almost exactly the arguments made for the Human Genome Project as a whole. It is Darwinism writ large, small, or on any scale you can imagine. Genomics corporations claim to shrink evolutionary scales of a billion years to the order (in both senses of the term) of a day (see Figure 3). For states, corporations, genomicists, technoscientific tools, and biomaterials alike, it has become a matter of survival of the fastest.

Notes

1. Cf. Paul Virilio, *Lost Dimension*, trans. Daniel Moshenberg (New York: Semiotext(e), 1991 [1986]). See also Michael Fortun, "Designed for Speed: Accelerating Genomics," in *The Practices of Human Genetics*, ed. Fortun and Everett Mendelsohn, Sociology of the Sciences Yearbook 19 (Dordrecht: Kluwer, 1996).

2. See, e.g., Daniel J. Kevles and Leroy Hood, "Reflections," in *The Code of Codes: Scientific and Social Issues in the Human Genome Project*, ed. Kevles and Hood (Cambridge, Mass.: Harvard University Press, 1992), pp. 300–328, esp. 306–7.

3. Robert G. Martin, "Why Do the Human Genome Project?" *NIHAA (National Institutes of Health Alumni Association) Update* 1, 2 (Autumn 1989), 4–5; and Bernard Davis, "What's the Big Hurry?—Thoughts on the Human Genome Project," ibid., pp. 6–7, on p. 6.

4. Quoted in Larry Casalino, "Decoding the Human Genome Project: An Interview with Evelyn Fox Keller," *Socialist Review* 91, 2 (1991), 111–28, on p. 114.

5. For an elaboration on why the Human Genome Project—which encompassed more than the human and is focused less on genomes and more on the production of tools for manipulating genomes—should be renamed "the genomics project," see Michael Fortun, "Mapping and Making Genes and Histories: The Genomics Project in the U.S., 1980–1990," Ph.D. dissertation, History of Science, Harvard University, 1993.

6. Gilles Deleuze and Felix Guattari, *A Thousand Plateaus: Capitalism and Schizophrenia* (Minneapolis: University of Minnesota Press, 1987), p. 17.

7. See Joan H. Fujimura, "Crafting Science: Standardized Packages, Boundary Objects, and 'Translation,' " in *Science as Practice and Culture*, ed. Andrew Pickering (Chicago: University of Chicago Press, 1992), pp. 168–211; and Fujimura, *Crafting Science, Transforming Biology: The Case of Oncogene Research* (Cambridge, Mass.: Harvard University Press, 1996).

8. Kathleen Jordan and Michael Lynch, "The Sociology of a Genetic Engineering Technique: Ritual and Rationality in the Performance of the Plasmid Prep," in *The*

Right Tools for the Job: At Work in Twentieth Century Life Sciences, ed. Adele E. Clarke and Joan H. Fujimura (Princeton, N.J.: Princeton University Press, 1992), pp. 77–114.

9. Eric S. Lander and David Botstein, "Mapping Complex Genetic Traits in Humans: New Methods Using a Complete RFLP Linkage Map," *Cold Spring Harbor Symposia on Quantitative Biology* 51 (1986), 49–62, on p. 49 (emphasis added).

10. See Fortun, "Mapping and Making Genes and Histories" (note 5), chap. 3, for an extended discussion of the development of this particular area of genomics; see also Jerry E. Bishop and Michael Waldholz, *Genome: The Story of the Most Astonishing Scientific Adventurre of Our Time* (New York: Simon and Schuster, 1990), pp. 49–80.

11. Ray White, "In Search of DNA Polymorphism in Humans," in *Cancer Incidence in Defined Populations*, ed. John Cairns, Joseph L. Lyon, and Mark Skolnick, Banbury Report 4 (Cold Spring Harbor, N.Y.: Cold Spring Harbor Laboratory, 1980), pp. 409–20, on p. 419.

12. Fred Gebhart, "Bios Labs Aims to Be Top Supplier to Human Genome Project Researchers," *Genetic Engineering News* (1 June 1992), 20.

13. Paul Virilio, *Speed and Politics*, trans. Mark Polizzotti (New York: Semiotext(e), 1986), pp. 46–47.

14. Charles Cantor, address at Human Genome I conference, 2–4 Oct. 1989, San Diego, California (my transcript from tape).

15. Quoted in U.S. Congress, Office of Technology Assessment,"Issues of Collaboration for Genome Projects: Transcript of a Workshop Held June 26, 1987" (Springfield, Va: National Technical Information Service, U.S. Department of Commerce, 1988) (PB 88–162797), pp. 113, 115.

16. As a related note, one of Aaron Wildavsky's conclusions about the politics of genetic engineering is that "while efforts at regulatory control are likely to be tried, they are even more likely to fail," primarily because "the speed of discovery is so great that regulation cannot keep up with it. There are too many holes in too many dikes for a regulator to keep a finger in all or more [sic.] of them." Aaron Wildavsky, "Public Policy," in *The Genetic Revolution: Scientific Prospects and Public Perceptions*, ed. Bernard D. Davis (Baltimore: Johns Hopkins University Press, 1991), pp. 77–104, on p. 96.

17. Mark Skolnick, H.F. Willard, and L.A. Menlove, "Report of the Committee on Human Gene Mapping by Recombinant DNA Techniques," *Cytogenetics and Cell Genetics* 37 (1983), 210–73; and Robert S. Sparkes, "Human Gene Mapping Workshop VII," *American Journal of Human Genetics* 35 (1983), 1334–35.

18. Albert de la Chapelle, "The 1985 Human Gene Map and Human Gene Mapping in 1985," *Cytogenetics and Cell Genetics* 40 (1985), 1–7, on pp. 1, 6.

19. H.F. Willard et al., "Report of the Committee on Human Gene Mapping by Recombinant DNA Techniques," *Cytogenetics and Cell Genetics* 40 (1985), 360–489, on pp. 360, 363.

20. Frank Ruddle and Kenneth K. Kidd, "The Human Gene Mapping Workshops in Transition," *Cytogenetics and Cell Genetics* 51 (1989), 1–2.

21. Mark Skolnick and Uta Francke, "Report of the Committee on Human Gene Mapping by Recombinant DNA Techniques," *Cytogenetics and Cell Genetics* 32 (1982), 194–204, on p. 194.

22. G. Christopher Anderson, "Genome Database Booms as Journals Take the Hard Line," *The Scientist* (30 Oct. 1989), 4; and Christian Burks, "How Much Sequence Data the Databanks Will Be Processing in the Near Future," in *Biomolecular Data: A Resource in Transition*, ed. Rita Colwell (Oxford: Oxford University Press, 1989), pp. 17–26.

23. U.S. House of Representatives, *Biotechnology Competitiveness Act of 1988*, 1 Oct. 1988 (100th Cong., 2nd sess., H. Rept. 100–992), p. 1.

24. Martin, "Why Do the Human Genome Project"; and Davis,"What's the Big Hurry?" (both note 3).

25. See Fortun, "Mapping and Making Genes and Histories" (note 5), chap. 6, for further analysis of criticisms of the Human Genome Project in the United States.

26. U.S. Senate Committee on Energy and Natural Resources, *The Human Genome Project: Hearing Before the Subcommittee on Energy Research and Development*, 101st Cong., 1st sess., 11 July 1990, S. Hrg. 101–894, p. 119.

27. Quoted in Fortun, "Mapping and Making Genes and Histories" (note 5), chap. 2. Original tape recording deposited in the Human Genome Archives at the National Reference Center for Bioethics Literature, Georgetown University, Washington, D.C.

28. Senate Committee, *Human Genome Project* (note 26), p. 101.

29. For an analysis of capital's most recent (since 1973) bout of "time-space compression," see David Harvey, *The Condition of Postmodernity: An Enquiry into the Origins of Cultural Change* (Oxford: Blackwell, 1989), esp. pp. 260–308.

30. Electronic-mail communication between Charles DeLisi and David Smith, 30–31 Dec. 1985, "DOE Policies" file, Box BCD7, Human Genome Archive, National Reference Center for Bioethics Literature, Georgetown University, Washington, D.C.; see also Fortun, "Mapping and Making Genes and Histories" (note 5), pp. 88–91.

31. Quoted in Stephen S. Hall, "James Watson and the Search for Biology's 'Holy Grail,'" *Smithsonian* (Feb. 1990), 41–49, on p. 46.

32. Senate Committee, *Human Genome Project* (note 26), p. 33.

33. James D. Watson, "A Personal View of the Project," in *Code of Codes*, ed. Kevles and Hood (note 2), pp. 164–73, on 164–65 (emphasis added).

34. Paul Virilio, *Popular Defense and Ecological Struggles*, trans. Mark Polizzotti (New York: Semiotext(e), 1990), p. 87.

35. Barney G. Glaser and Anselm L. Strauss, *Status Passage* (Chicago: Aldine, 1971), p. 39 (emphasis added).

36. Robert Cook-Deegan, "The Human Genome Project: The Formation of Federal Policies in the United States, 1986–1990," in *Biomedical Politics*, ed. Kathi E. Hanna (Washington, D.C.: National Academy Press, 1991), pp. 99–168, on p. 161.

37. This is not a one-way relationship, i.e., money producing tools, but more the kind of "cycle of credit" first described by Bruno Latour and Steve Woolgar in 1979—see *Laboratory Life: The Construction of Scientific Facts* (Princeton, N.J.: Princeton University Press, 1986 [1979]), esp. pp. 187–233—or the kind of acentered machinic assemblage that Deleuze and Guattari describe, in which capital and tools (and other nodes of the assemblage as well) ceaselessly produce each other—see *A Thousand Plateaus* (note 6), esp. chap. 14, "1440: The Smooth and the Striated."

38. Robert Mullan Cook-Deegan, "Gene Quest: Science, Politics, and the Human Genome Project," 1991, typescript on deposit at Human Genome Archives, Georgetown University (note 27), and elsewhere, pp. 10.6–10.7.

39. National Research Council, *Mapping and Sequencing the Human Genome* (Washington, D.C.: National Academy Press, 1988), pp. 2–3.

40. Quoted in Jeffrey L. Fox, "Faster and More Accountable Genome Project," *Bio/Technology* 10, 1 (Jan. 1992), 120.

41. Earl Lane, "The Funding Ruckus," *Newsday*, 23 Oct. 1990, p. 21.

42. Earl Lane, personal communication, 18 March 1992.

43. See Leslie Roberts, "NIH Takes New Tack on Gene Mapping," *Science* 258 (1992), 1573.

44. Senate Committee, *Human Genome Project* (note 26), pp. 59–60.

45. Leslie Roberts, "The Rush to Publish." *Science* 258 (1991) 260–63, on p. 260.

46. Quoting from Gina Kolata, "Biologist's Speedy Gene Method Scares Peers But

Gains Backer," *New York Times*, 28 July 1992, p. C1. See also Edmund L. Andrews, "U.S. Seeks Patent on Genetic Codes, Setting Off Furor," *New York Times*, 21 Oct. 1991, pp. A1ff; Robin Eisner, "Biotechnology Community Mixed on NIH's Gene-Patenting Efforts" *The Scientist* (9 Dec. 1991), 1ff; Patrick D. Kelly, "Are Isolated Genes 'Useful'?" *Bio/Technology* 10, 2 (Feb. 1992), 52, 55; Leslie Roberts, "OSTP to Wade into Gene Patent Quagmire," *Science* 254 (1991), 1104–5; Roberts, "NIH Gene Patents, Round Two," *Science* 255 (1992), 912–13; Roberts, "Scientists Voice Their Opposition," *Science* 256 (1992), 1273–74; Roberts, "Rumors Fly over Rejection of NIH Claim," *Science* 257 (1992), 1855; and Scott Veggeberg, "Controversy Mounts over Gene Patenting Policy," *The Scientist* (27 April 1992), 1ff.

47. Kolata, "Biologist's Speedy Gene Method" (note 46), p. C10.

48. Eisner, "Biotechnology Community Mixed" (note 46), p. 10.

49. All quotations from Mark D. Adams et al., "Complementary DNA Sequencing: Expressed Sequence Tags and Human Genome Project," *Science* 252 (1991), 1651–56.

50. Kolata, "Biologist's Speedy Gene Method" (note 46), p. C1.

Data Access Policy in Genome Research

Stephen Hilgartner

A persistent source of tension in many areas of contemporary science is the issue of access to scientific data, broadly construed to include information, biological materials, and a variety of other resources. The issue raises basic questions about the proper conduct of science and about what is private and what is public in science; it is tightly intertwined with policy issues regarding university-industry relations and intellectual property.[1] Conflicts about data access also raise practical concerns for agencies that fund science, because scientific secrecy or disputes about ownership of data can slow the pace of research.

This essay examines several policies aimed at managing the tensions surrounding access to data that funders of genome research have adopted. Much of this research is conducted under the rubric of the Human Genome Project (HGP), an international effort to develop genomic technology and to map and sequence the genomes of the human and several other organisms. The HGP is a useful research site for this inquiry for several reasons. Debate about access to data has been common in molecular genetics and genome research, very competitive areas of science with commercial applications and major grants at stake. Moreover, data access has been a salient issue for HGP policy makers, who face the task of completing this long-term program of targeted research without overrunning budgets.

Here I explore three of these policies, first introducing the HGP, then describing an analytic framework for examining access practices and policies aimed at influencing them.[2]

The Human Genome Project

The HGP, a loosely coordinated international effort, began to take shape at the end of the 1980s. The central aims of the project are to improve technology for analyzing large genomes, and to map and sequence the human genome and those of a handful of "model" organisms.[3] In the United States, the project is being funded and managed by two federal agencies, the National Institutes of Health (NIH) and the Department of Energy (DOE), which have set the goal of completing the human sequence by the year 2005 at a cost of $3 billion. Genome programs of varying sizes and structures were also established in Europe, Canada, and Japan.

Policy discussion about access to genome project data began even before the U.S. genome program was officially initiated. In 1988 an influential report of the U.S. National Research Council (NRC), *Mapping and Sequencing the Human Genome*, argued that "access to all sequences and materials generated by these publicly funded projects should and even must be made freely available." The project would "require an unprecedented sharing of materials among the laboratories involved."

> The human genome project will differ from traditional biological research in its greater requirement for sharing materials among laboratories. For example, many laboratories will contribute DNA clones to an ordered DNA clone collection. These clones must be centrally indexed. Free access to the collected clones will be necessary for other laboratories engaged in mapping efforts and will help to prevent a needless duplication of effort. Such clones will also provide a source of DNA to be sequenced as well as many DNA probes for researchers seeking human disease genes.[4]

The NRC report, and other statements like it, reflect the general commitment of the leadership of the HGP to a broad philosophy of widespread access to data. But to move from an abstract commitment to its actual implementation poses significant challenges, especially when goals are stated in vague terms that allow for divergent interpretations. For one thing, a broad philosophy of "free access" does not begin to answer potentially contentious questions about exactly which data should be made available to whom, when, and under what terms. These matters are not trivial administrative details, but are the very essence of the policies and practices that constitute greater or lesser freedom of access. In addition, a broad philosophy leaves open the question of precisely how policies promoting free access should be structured and enforced. That these issues have at times provoked disagreement is not surprising, in light of competition for funding, priority, and intellectual property rights.

Below, I explore some policies that genome project policy makers in the United States and Europe formulated for the express purpose of influencing access practices. I first introduce the concept of data streams, then examine three examples of access policies in genome research.

Data Streams and Access Policy

An analysis of policy making about data access must be rooted in a framework for understanding access practices more generally. The most familiar such framework relies on Robert K. Merton's theory of the normative structure of the scientific community. This approach is exemplified in Warren Hagstrom's analysis of scientific communication as "gift exchange" and Katherine W. McCain's more recent work on the dissemination of research-related information in genetics.[5] However, the gift exchange approach is too far removed from the material practices and day-to-day life of the laboratory to shed much light on how scientists control access to data through their routine activities. Working with Sherry Brandt-Rauf, I developed a framework, rooted in the findings of ethnographic studies of scientific practice, based on the analysis of "data streams."[6]

Building on constructivist theory, the data stream perspective treats the category "data" as problematic. Rather than assuming that data consist of well-defined objects, neatly packaged and clearly bounded, data are seen as embedded in evolving streams of scientific production, or data streams. These data streams have a number of important characteristics. First, they are composed of a heterogeneous mix of components, including written inscriptions, biological materials, computer software, laboratory techniques, and tacit knowledge. These components differ in familiarity and availability. Some may be completely mundane; some may be specialized, but widely available scientific paraphernalia; some may be unique to the laboratory that possesses them. The components also vary in factual status. Some are considered highly uncertain and some are widely believed to be reliable.[7]

Second, data streams consist of complex assemblages that weave together the heterogeneous components that constitute them. These assemblages continuously evolve during scientific production. As laboratory work proceeds, scientists process and reprocess materials and create inscriptions, such as dark bands on X-ray film, that are converted into numbers, subjected to further analysis, and presented in tables and graphs.[8] Data streams thus consist of chains of inscriptions, materials, and other entities; they are evolving assemblages that, generally speaking, grow more complex and more densely interconnected as scientific work proceeds.

These properties have important implications for access practices. For example, different elements of these heterogeneous assemblages can be put to different uses, making certain portions of them particularly valuable. More fundamentally, "the fact that data streams are composed of chains of products suggests the advantages of conceiving of data streams as continuous phenomena, a move that shifts the level of analysis from the individual end-product to the stream as a whole."[9] This disrupts the very notion of an "end-product" and brings out a central question of the data stream perspective. Given that data streams emerge as continuous, evolving assemblages, what portions of these streams get provided to whom, when, and under what terms and conditions?

Third, the data stream perspective highlights the strategic considerations involved in decision making about access. The perspective can be combined with various models of human decision making, but regardless of the model used, it assumes that scientists seek to successfully exploit their data. It also emphasizes the wide array of possible methods for exploiting valuable data—methods that include publishing, patenting, selling, bartering, and a variety of other types of transactions. These transactions range from those that result in widespread access to those that keep the data completely confidential.[10] Another important class of transactions provides strictly controlled, targeted access—for example, when scientists exchange data as part of the terms of a collaboration. These transactions may involve complex, ongoing negotiations.[11]

The data stream perspective also stresses the issue of timing. Precisely *when* is access provided?[12] This question emerges in many contexts, and its strategic implications are deepened by two kinds of considerations. First, scientists must ask themselves whether the data are sufficiently reliable that it is appropriate to release them. Second, researchers must consider when to engage in which types of transactions: scientists often can deploy different access mechanisms in succession (e.g., by first providing carefully targeted access in exchange for other resources, then later providing more widespread access through publication). To further complicate the strategic issues surrounding timing, issues of quality control and access control can interact. Providing data of uncertain status may pose risks to reputation, for example, and efforts to verify results may be a reason, or serve as an excuse, for delay.

Scientists are not the only ones who try to control access to data: science-policy makers have their own reasons for—and methods of—influencing scientists' access practices. By *science-policy makers* I refer to representatives of governmental funding agencies (or private foundations that fund significant amounts of research) and to scientists (and others) who advise them. Their specific aims may include maintaining productivity in their research domain, getting data and resources into use, reducing duplication of effort, avoiding disputes over access and ownership that might disrupt research, rewarding researchers who exhibit "good" behavior, and preserving their legitimacy as science-policy makers.

In their attempts to influence access practices, policy makers use a wide range of techniques; from admonishing researchers informally to share data to imposing formal rules about data access on grantees. Scientists, however, cannot be assumed simply to acquiesce to policy makers' desires: strategic considerations and collective definitions of appropriate conduct mediate their responses to particular policies.[13] Thus the actions of policy makers should not be viewed in isolation, but as part of a dynamic interplay of influence and resistance that shapes the ways particular policies affect actual practices. With this analytic framework in mind, let us turn to three empirical examples.

Single Chromosome Workshops

The first example concerns the Single Chromosome Workshops (SCWs), a series of international meetings, each focused on a different human chromosome. The SCWs brought together researchers from around the world to exchange data and develop a state-of-the-art map of the chromosome in question. For the funding agencies that supported these workshops, sharing and comparing data and materials constituted a central goal, essential to the construction of a map based on the best data available.[14] However, after some participants were less than forthcoming with their data at several early SCWs, genome project policy makers began seriously discussing formal data-sharing requirements for the workshops.[15]

Data access became a salient issue for the SCWs because of the kinds of data typically presented at them. Two main lines of research were strongly represented. Many participants were active "gene hunters"—scientists searching for particular human genes, most often those implicated in disease. Other participants were "chromosome mappers," working to construct global maps of entire chromosomes. Despite their different goals, both lines of research involve creating maps, and each can contribute useful data to the other. Although gene hunters typically construct a high-resolution map in the region where the gene is believed to be located, while chromosome mappers create low-resolution maps of entire chromosomes, the two types of maps can share landmarks, so that data originally developed for one can be incorporated in the other. Just as the data streams of these lines of research can overlap, so too can the scientists: a number of chromosome mappers are also active gene hunters.

Searching for human disease genes is a fast-paced, competitive area of research. Identifying a gene implicated in a well-known disease conveys prestige and can result in the award of valuable patents. In most cases, several research teams end up targeting each major disease gene; but just *one* group can win the "race." To further raise the stakes, in the early 1990s gene hunters could expect to devote a number of years to the search for a single gene, although today improved technology and expanded collections of map and sequence data have significantly shortened the time required.

Besides the competitive pressure in this high-risk, high-reward area of research, the structure of their data streams discourages gene hunters from providing widespread access. Using the techniques of the early 1990s, searching for a gene entails creating a detailed map in the region where the gene is believed to be located. In many cases, any map data that a researcher publishes can immediately be employed by competitors, who can combine the new data with their own to create even better maps.[16] Thus when gene mappers publish data or make materials available, they often face the risk that competitors will instantly catch up or even get ahead. Gene hunters at the SCWs thus faced contradictory pressures: an incentive to share data, because it would improve the global map of the chromosome and significantly speed progress in human genetics; but a disincentive to release data that might allow their direct competitors—who were often present at the workshops—to leapfrog ahead.

How scientists managed these tensions varied considerably. Not surprisingly, gene hunters tended to control carefully the disposition of their data streams. Many were quite forthcoming about sharing data and resources. A few earned reputations for being stingy. But many or most made careful strategic assessments about which portions of their data streams to distribute and which to keep private. One prevailing strategy is what we have termed *delayed release*—maintaining a significant gap between the data in the lab and the data presented publicly.[17] This strategy allows scientists to engage in mutually beneficial exchanges or to demonstrate progress to colleagues, while controlling the risk that competitors will seize the lead. If the delay is sufficient, the risk will be small.

Although the policy makers responsible for the SCWs recognized that gene hunters practiced delayed release, they were most concerned about several other practices:

- Some participants presented data at the meetings and then failed to submit them to the Genome Data Base (GDB), which would make them accessible worldwide via the Internet.
- Some attended the meetings but behaved as what were derisively referred to as "spectators"—they collected data but contributed nothing of their own.
- Some participants engaged in "data isolation"—they provided access to isolated portions of a data stream that had been disassembled so that it could not be readily used by other laboratories.[18] For example, a researcher might present results in talks or posters, but decline to release the clones needed to build on or verify the results.

In an effort to curb these practices, the agencies that sponsored the SCWs amended the guidelines for the meetings to require that "any data and materials that are presented as an abstract or presented at the workshop are automatically considered to be immediately publicly available."

This policy not only eliminated spectators and encouraged submission of data to GDB, but it curbed data isolation. It did not necessarily lead everyone to become more open with data, however. Some decided not even to mention work that they had completed so that they could not be required to make data public. One gene hunter told me that she had carefully avoided referring to several important results in her talk because she did not want to be compelled to provide the relevant biological materials to her competitors, who were present at the meeting. The SCW policy on sharing data and materials could thus eliminate some methods of restricting access, but researchers could still resort to a variety of other strategies. These limitations, however, should not be interpreted simply as a defect in the policy, but as a reflection of the constraints that data access policy must inevitably face.

Sequence-Tagged Sites

My second example, the decision of the American genome program to emphasize a particular type of mapping landmark—the sequence-tagged site, or STS—represents a policy intended to achieve several goals, including influencing access practices. In 1989 a scientific advisory group working on the first five-year plan for the American genome program endorsed the STS concept.[19] The NIH and DOE quickly adopted STSs, and the five-year plan defined the goals for genome mapping in those terms. This move, in effect, required genome mapping laboratories to use STSs, and STSs became the "centerpiece" of the American genome mapping effort.[20] The policy makers involved favored STSs for a variety of interrelated reasons, but among the significant ones was the notion that using the new landmarks would make it harder for researchers to monopolize access to map data.

How could insisting that mappers use a particular kind of mapping landmark affect access to data? The structure of data streams in genome mapping suggests the answer. In any form of cartography—whether the scale is galactic or molecular—maps are constructed by defining the spatial relations among a set of recognizable landmarks. In genome mapping, maps are constructed by locating distinct landmarks on the genome to use as points of reference. As mapping technology evolved, researchers developed several different kinds of maps (including contig maps, radiation hybrid maps, and restriction maps), each based on a different type of landmark. Before STSs, all these landmarks relied on clones—bacteria or yeast into which the DNA fragments under study have been inserted. In order to test a DNA sample for the presence of one of these landmarks, one had to have access to the clone on which it was based. Put otherwise, to use such a map, one needed more than simply the written description of it; one also needed the biological material used to construct it. In effect, each landmark on these maps was an assemblage of inscription and material, *both* of which had to circulate among laboratories to make the maps widely available to the scientific community.

This simple fact—that inscriptions and clones both must circulate—had profound implications for the HGP. Whereas inscriptions can be readily disseminated using print and electronic media, clones must be stored and shipped in the appropriate culture media. Clones can only be replicated in laboratories; contamination must be avoided; and they must be carefully packed for shipping. Moreover, genome maps involve large numbers of clones and the task of handling them is labor intensive.

For clone-based maps, the requirement that clones and inscriptions circulate through different channels complicates data sharing. Distributing clones creates logistical problems for those who do want to provide access to data. If one scientist releases the inscriptions describing a map, other scientists who seek to use the map have to obtain the clones on which it is based. Even for a large mapping lab, the prospect of having to manage thousands of requests for clones was daunting. In the long run, HGP was expected to generate hundreds of thousands of clones, which would have to be stored in an expensive, highly automated repository, complete with a mail-order house.[21]

The need for special arrangements to circulate clones also makes it easier to restrict access to data. Researchers who wish to delay access can process requests for clones slowly. They can release clone-based maps but decline to provide clones until some later date—explaining, for example, that their work is still preliminary or that they will release the clones once they publish a more complete, "final" map. These excuses make it impossible for the requester to determine whether the intent of the delay is to ensure quality or to delay access. The full range of rhetoric and techniques for delaying access to clones need not be delineated here; suffice it to say that the logistics of clone distribution can create strategic opportunities for using data isolation to restrict access.

STSs avoid these problems because, unlike all previous mapping landmarks, they depend not on clones but on the polymerase chain reaction (PCR), an in vitro method for making copies of specifically targeted DNA fragments without cloning. PCR allows STSs to be described completely as a written inscription, such as

TCTGTTTTGGATAACAATGCC ACAGCTTCCCCATCTACTCC,

which identifies an STS on human chromosome 1. To test a DNA sample for this STS, all one needs is this inscription and a standard PCR set-up. As the scientists who outlined the STS concept put it, "no access to the biological materials that led to the definition or mapping of an STS is required by a scientist wishing to assay a DNA sample for its presence."[22]

This trait in part explains the enthusiasm for STSs of the original advisory group to the NIH and DOE.[23] By requiring grantees to describe their maps in terms of STSs, the funding agencies could accelerate the distribution of maps to the entire scientific community. Researchers who published STSs or placed them in electronic databases would have no way to prevent others from using and building on their maps. Moreover, the large clone repository was obviated, greatly simplifying the logistics of data sharing. In short, the policy of requiring genome mappers to use STSs as landmarks built mechanisms for speeding access to data into the very means of scientific production.

The European Yeast Sequencing Program

The third example comes from the world of sequencing—that is, the determination of the order of nucleotides in a DNA sample. One the most ambitious genomic sequencing efforts of the early 1990s was the Commission of the European Communities (EC) program of large-scale sequencing of the genome of the yeast *Saccharomyces cerevisiae*. The EC selected yeast for several reasons. Yeast is an important model organism in a range of biological research, and it serves as a key "tool" for a community of yeast researchers.[24] *S. cerevisiae* is also well-suited to studies of gene function. Another advantage of yeast is that its genome is relatively small, which led EC policy makers to believe that it could be sequenced using the technology anticipated in the early 1990s. Sequencing yeast also won the support of several major industries, including pharmaceuticals, chemicals, food, and brewing, for which the organism has economic importance. Finally, sequencing yeast was not expected to encounter ethical objections, whereas a highly visible sequencing project aimed at *human* DNA would face political opposition in Germany, the largest European country, given the history of eugenics under the Third Reich.

In 1989 the EC began a pilot project to sequence yeast chromosome III, and it completed its 315,000-base-pair sequence in 1991. That same year, the sequencing program—which was widely perceived as a success—was expanded to include chromosomes II and XI. Later, still more of the sixteen yeast chromosomes were targeted (including chromosomes II, VII, X, XIV, XV, and part of XII and IV). Sequencing was performed chromosome by chromosome. Each chromosome was managed as a separate project, all of which followed the organizational model developed for the chromosome III pilot project. The EC yeast program was widely perceived as successful, and it made a significant contribution to the development of a concerted effort—involving laboratories in Europe, North America, and Japan—to obtain the complete sequence of the yeast genome. The achievement of this goal was announced at an April 1996 news conference, with the European program contributing the largest amount of sequence data.

In principle, a major sequencing project can be organized in a variety of ways, ranging from conducting it in a single, large sequencing center to orchestrating a collaboration involving many geographically dispersed laboratories. For several reasons the EC opted for a "network" model that would link existing laboratories into a coordinated effort, although critics—especially in the United States—contended that the model could not achieve the economies of scale of a large center. The EC science-policy makers believed that setting up a large sequencing center would require lead time to hire staff and build the laboratory infrastructure, whereas a network model that used existing laboratories could be started up more quickly. They considered a network model more "robust" because it did not require investing all one's resources in a single facility: If a few laboratories failed to produce usable data, the project as a whole could still succeed.[25]

The network model also had significant political advantages, given the EC's international composition. Since the network would be distributed throughout Europe, researchers in many European countries would gain experience with large-scale sequencing, and this would facilitate the economic development that is a central goal of European integration.[26] Moreover, housing the project in a single central facility would entail deciding in which country it should be located, whereas the network model would circumvent this potentially divisive question.

The task of building and running a network of sequencing laboratories spread across Europe is by no means a trivial one. Qualified laboratories must be identified and induced to participate. A system for coordinating the work must be devised. Tasks must be assigned, information and materials must be distributed, and means for controlling the quality of data must be set up. In addition to managing the flow of data and work, the roles of the various participants must be structured so as to prevent conflicts from disrupting the network.

Strategies for managing tensions about access and ownership of data were a fundamental part of the design of the network model. To ensure that the data stream would develop in an orderly manner, without ownership disputes or unnecessary duplication of effort, the EC established clearly stated rules about the production, ownership, and exchange of the elements of the data stream.

Once again, examining the structure of data streams helps us understand these rules. In most large-scale genomic sequencing projects during the early 1990s, data streams took the following general form, although the details of specific projects differed in significant ways. At the front of the data stream, the work begins with the production of a detailed physical map of the region to be sequenced. A key part of this map is an ordered collection of clones spanning the region, generally cosmid clones in the EC yeast program. These ordered cosmids then serve as the material to be sequenced by the sequencing labs. Each cosmid is sequenced separately and its sequence is placed in a computer database. Next is sequence assembly. The computer identifies the overlapping bits of sequence that appear at points where the cosmids overlap, thus merging the sequence into a continuous array. The computer also performs various quality checks on the data; some DNA may be resequenced to establish accuracy or resolve discrepancies. Eventually, the entire sequence, spanning the target region in an unbroken array, is completed.

Two additional points about the structure of this data stream are crucial.

First, with the technology of the early 1990s, the sequencing of cosmid-scale DNA segments was (and remains today) by far the most laborious part of the work. Second, this sequencing can be performed in parallel, allowing it to be divided among many groups, each of which can sequence a different set of cosmid clones. A large sequencing center might house a number of such groups; alternatively, each group might be housed in a different laboratory. In either case, these groups can in principle operate quite independently—and even use different sequencing methods—so long as they submit their data to a central database for assembly of the final sequence.

The EC network model exploited the fact that sequencing can be done in parallel. This labor-intensive task was divided among several dozen sequencing labs, bound together by a set of contracts, rules, and understandings that governed the division of labor, the flow of data and materials, and the ownership of the data stream as it evolved over time. (To give a sense of scale, the pilot project to sequence chromosome III involved thirty-five laboratories drawn from ten European countries.) The sequencing labs were awarded contracts to sequence particular segments of the chromosome, and they were paid per base pair of data produced. These labs enjoyed considerable independence, but a central administrative structure coordinated their work. Each chromosome project had a "DNA Coordinator," who was responsible for physical mapping and supplying clones to the sequencing labs. One "Informatics Coordinator" served all chromosomes, managing a database that stored the data produced by the sequencing labs and performing computerized data checking, sequence assembly, and preliminary analysis. After looking for genes and other biologically interesting features, the Informatics Coordinator provided the results to the lab that supplied the sequence and to the DNA Coordinator. The sequencing lab could also perform sequence analysis and experimental studies of the function of any genes identified, and engage collaborators in this effort. Close interaction between the DNA and Informatics Coordinators facilitated ongoing management.

The EC program offered incentives to interest competent laboratories in participating, including payments generous enough to let sequencing labs make a "profit." For academic labs interested in yeast biology, the extra income might fund other work, including studies of the biological function of the DNA they had sequenced. The arrangement also attracted some industrial laboratories looking to develop sequencing experience so they later could offer DNA sequencing as a commercial service. Beyond financial incentives, participating researchers benefited from early access to valuable data on the yeast genome. The network granted all participants rights to certain portions of the data stream. An explicit set of rules designated some data as confidential and specified who had the authority to publish what, when, and under whose authorship. (Adjustments in the rules, which were changed slightly as experience accumulated, are not addressed here.) At the front end of the data stream, for example, the DNA Coordinator had the right to publish the map of the chromosome. The map was provided to all sequencing labs, who were permitted to use it internally, but not to publish it or otherwise distribute it.

Perhaps most interesting, the rules on the disposition of the data stream granted what the scientists called "property" rights to regions of the genome. At the outset of sequencing, the DNA Coordinator divided up the chromosome, assigning each sequencing lab a region to analyze. These assignments conveyed "ownership rights"—again, the scientists' term—in the region.[27] These rights had several dimensions:

- The region became the participant's "property." The sequencing lab was entitled to sequence the region, and it could not be claimed by other members of the network.
- The lab was entitled to publish, patent, or otherwise distribute the sequence data from the region. Before distributing the sequence, however, the lab was required to submit it to the Informatics Coordinator, who would keep it confidential.
- Ownership rights entitled sequencing labs to engage in collaborations with scientists of their choosing to study the biological function of the DNA they sequenced.
- Ownership rights could be canceled if a lab failed to produce sequence data of acceptable quality on schedule.
- Ownership was of limited duration. As soon as the sequence of the entire chromosome was completed, the ownership rights of all participants would be canceled. The entire sequence would be published (with all the participants as authors) and made public in a database operated by the European Molecular Biology Laboratory (EMBL), which is accessible worldwide via the Internet.

The ownership policy played an important role in coordinating the network. By parceling out regions of the chromosome, the European yeast program avoided duplication of effort and prevented "races" in which two or more laboratories competed to sequence the same region. The ownership policy also encouraged sequencing labs to perform well: because ownership could be canceled, the rules spurred them to meet the network's standards regarding quality and speed; and the rules motivated them to publish their sequence rapidly, while they still retained ownership rights.

At another level, the rules about ownership helped to manage the interaction between quality control and access control. Under the rules, any data that a sequencing lab publishes prior to the completion of the entire chromosome are released under the authorship of that lab, which is solely responsible for their quality. Only later, after the network has performed a series of verification steps designed to check the quality of the sequence, are the data considered "final" sequence, backed by the authority of the network.[28] This arrangement was intended to protect the credibility of the network as a whole from being compromised, should a laboratory publish a sequence containing many errors.

The rules of the EC yeast network represent a relatively comprehensive system for governing the disposition of a data stream produced collectively by a large, distributed group of collaborators. The relatively strong control that the contract mechanism offers allowed the EC to establish a social order instantiated in a division of labor that was linked to—and reinforced by—a set of rules that defined the boundaries used in apportioning the data stream. Some of these boundaries were spatial (i.e., regarding regions of the genome); some were temporal (private now, public when the chromosome is completed); and some referred to categories of data (e.g., map versus sequence data, preliminary versus final data).

This social order, however, existed only within what was clearly the most important boundary of all: the one that defined the network itself. The rules, after all, only applied to network participants, who had to accept them to join. Within the network, potential tensions surrounding data access were successfully managed, and the European program was able to produce a large volume of sequence data. From outside, however, there were recurring complaints that the network was too slow in releasing results to the wider scientific community. Yeast biologists around the world eagerly awaited the data; whenever the network released new sequence, many labs accessed it within hours of its being made available on the EMBL computer. Not surprisingly, some scientists grew impatient and called for more "timely" releases. Members of the EC yeast network countered that a period must be allowed for verification and for the scientists who did the sequencing, who sweated hard to produce the data, to perform a preliminary analysis. As one DNA Coordinator explained:

> We need to keep people motivated. If they don't get a benefit from what they do— It is very easy to be a retired person and sit on the computer and extract a nice idea out of sequences if you don't spend your time running the [sequencing] gels. It's very easy and it's very unfair, I would say, for the people who are scientists running the gels.

Disagreement about the timing of releases is a persistent theme in debate about access policy in genome mapping and sequencing. Despite its success in managing access issues internally, the EC yeast network, like many other research groups, was unable to escape debate about what constitutes a reasonable period of delay for ensuring accuracy or for rewarding data producers with a short-term monopoly on access.

Summary and Discussion

The three examples of access policies examined above from a data stream perspective illustrate the salience of access policy in the genome project and in the world of genome research generally. The analysis underlines the importance of taking account of the competitive pressures scientists face, the patterns of incentives that may influence their behavior, the strategic options at their disposal, and the structure of the relevant data streams.

The three policies all arose in different contexts and had different goals, so they cannot be compared to determine which was the most "successful" or which represents the most effective approach to access policy. Comparing these cases, however, clearly demonstrates the wide variety of ways by which access policy can be insinuated into the sociotechnical fabric of science. Some, like the data-sharing requirements for SCWs, consist of simple injunctions that simply and transparently define appropriate conduct. Access policy can also be built directly into the "technical" means of scientific production, as the use of STSs illustrates. The yeast sequencing network represents yet another, very different means of establishing access policy: the EC created an organizational structure, backed by a rather detailed set of rules and contracts, that specified the roles of participants in the production and disposition of the data stream.

These cases also make clear the limits of access policy. Although all the policies considered above influenced access practices, they were effective only within relatively narrow domains. The ownership policy of the European yeast sequencing network applied only to participants in the program. The data-sharing requirements of the SCWs only addressed these specific meetings, and they provided incomplete control over scientists' practices. The policy of using STSs to represent map data was widely seen as a successful means of providing access to mapping landmarks, but it applied only to these landmarks and left many other critical resources untouched. Clearly, it would be naive to imagine that a single policy could simply "resolve" the problems surrounding access to genome data: access issues reflect tensions—between competition and cooperation, between public and private—that are deeply embedded in the structure of contemporary research systems.

In appraising the significance of access policies, one must consider not only their immediate practical implications, but also their symbolic importance in the ongoing negotiations that constitute the meaning of such terms as "free access." After all, these policies not only implement particular rules, they also instantiate particular definitions of appropriate scientific practice. The data-sharing requirements of the SCWs, for example, place data isolation outside the boundaries of propriety. The STS policy embodies a commitment to making a particular kind of knowledge—genome maps—into public knowledge. And the rules governing the European yeast network express a particular vision of the rights and obligations of scientists engaged in collaborative projects. Access policies thus have symbolic and rhetorical significance in the continuing debates that are reshaping the boundary between public and private in late twentieth-century research systems. Changes in the contours of that boundary raise critical issues not only in genomics, but in many technical fields. Consequently, further research on access policy and practices promises to be a fruitful area for social studies of science.

Research for this article was funded in part by the National Center for Human Genome Research of the National Institutes of Health and by the National Science Foundation.

Notes

1. For a few examples from this broad literature, see Rebecca S. Eisenberg, "Property Rights and the Norms of Science in Biotechnology Research," *Yale Law Journal* 97 (1983), 177–231; Henry Etzkowitz, "Entrepreneurial Science in the Academy: A Case of the Transformation of Norms," *Social Problems* 36 (1989), 14–29; Martin Kenney, *Biotechnology: The University-Industrial Complex* (New Haven, Conn.: Yale University Press, 1986); Sheldon Krimsky, *Biotechnics and Society: The Rise of Industrial Genetics* (New York: Praeger, 1991); Michael Mackenzie, Peter Keating, and Alberto Cambrosio, "Patents and Free Scientific Information in Biotechnology: Making Monoclonal Antibodies Proprietary," *Science, Technology, and Human Values* 15 (1990), 65–83; National Academy of Sciences, *Sharing Research Data*, ed. Stephen E. Fienberg, Margaret E. Martin, and Miron L. Straf, Committee on National Statistics, National Research Council (Washington, D.C.: National Academy Press, 1985); National Academy of Sciences, *Responsible Science: Ensuring the Integrity of the Research Process* (Washington, D.C.: National Academy Press, 1992); Dorothy Nelkin, *Science as Intellectual Property: Who Controls Research* (Washington, D.C.: American Association for the Advancement of Science, 1984); Richard R. Nelson, "What Is Private and What Is Public About Technology," *Science, Technology and Human Values* 14 (1989), 229–41; and Vivian Weil and John W. Snapper, eds., *Owning Scientific and Technical Information: Value and Ethical Issues* (New Brunswick, N.J.: Rutgers University Press, 1989).

2. This article is drawn from a long-term study of the HGP. Besides the sources in the notes, it relies on participant observation at genome project meetings and interviews with American and European genome scientists and policymakers. Respondents were promised anonymity.

3. These "model" organisms include *E. coli*, yeast, *C. elegans*, *Drosophila*, and the mouse. On the HGP see, e.g., Robert Cook-Deegan, *The Gene Wars: Science, Politics, and the Human Genome* (New York: W.W. Norton, 1994); Stephen Hilgartner, "The Human Genome Project," in *Handbook in Science, Technology and Society*, ed. Sheila Jasanoff et al. (Newbury Park, Calif.: Sage, 1994), pp. 302–15; Bertrand Jordan, *Travelling Around the Human Genome* (Montrouge, France: John Libbey Eurotext, 1993); Daniel J. Kevles and Leroy Hood, eds., *The Code of Codes: Scientific and Social Issues in the Human Genome Project* (Cambridge, Mass.: Harvard University Press, 1992); and the article by Michael Fortun in this volume.

4. National Research Council, *Mapping and Sequencing the Human Genome* (Washington, D.C.: National Academy Press, 1988), pp. 8, 76.

5. Robert K. Merton, "The Normative Structure of Science," in *The Sociology of Science* (1942), ed. Merton and Norman W. Storer (Chicago: University of Chicago Press, 1973); Warren O. Hagstrom, *The Scientific Community* (New York: Basic Books, 1965); and Katherine W. McCain, "Communication, Competition, and Secrecy: The Production and Dissemination of Research-Related Information in Genetics," *Science, Technology, and Human Values* 16 (1991), 491–516.

6. Stephen Hilgartner and Sherry I. Brandt-Rauf, "Data Access, Ownership, and Control: Toward Empirical Studies of Access Practices," *Knowledge: Creation, Diffusion, Utilization* 14 (1994), 355–72. For a critique of the gift exchange perspective, see Stephen Hilgartner and Sherry I. Brandt-Rauf, "Controlling Data and Resources: Access Practices in Molecular Genetics," in *A Productive Tension: University-Industry Research Collaborations*, ed. Paul A. David and W. Edward Steinmueller (Stanford, Calif.: Stanford University Press, forthcoming). See also Robert E. Kohler, *Lords of the Fly: Drosophila Genetics and the Experimental Life* (Chicago: University of Chicago Press, 1994) for a historical analysis of access practices among the early drosophilists that is sensitive to the material culture of the lab. Studies of scientific laboratories and practices that inform the data stream perspective include Bruno Latour and Steve Woolgar, *Laboratory Life: The Construction of Scientific Facts* (Beverly Hills, Calif.: Sage, 1979); Karin D. Knorr-Cetina, *The Manufacture of Knowledge: An Essay on the Constructivist and Contextual Nature of Science* (New York: Pergamon Press, 1981); Michael Lynch, *Art and Artifact in Laboratory Science* (Boston: Routledge and Kegan Paul, 1985); Harry M. Collins, *Changing Order: Replication and Induction in Scientific Practice* (Beverly Hills, Calif.: Sage, 1985); Joan H. Fujimura, "Constructing 'Do-able' Problems in Cancer Research," *Social Studies of Science* 17 (1987), 257–93; and Susan Leigh Star, *Regions of the Mind: Brain Research and the Quest for Scientific Certainty* (Stanford, Calif.: Stanford University Press, 1989).

7. Hilgartner and Brandt-Rauf, "Data Access, Ownership, and Control" (note 6), pp. 359–61. On the shifting epistemic status of the entities produced during scientific work see Latour and Woolgar, *Laboratory Life* (note 6); Knorr-Cetina, *Manufacture of Knowledge* (note 6); and Bruno Latour, *Science in Action: How to Follow Scientists and Engineers through Society* (Cambridge, Mass.: Harvard University Press, 1987).

8. On inscriptions see Latour and Woolgar, *Laboratory Life* (note 6); and Latour, *Science in Action* (note 7).

9. Hilgartner and Brandt-Rauf, "Data Access, Ownership, and Control" (note 6), p. 361.

10. Hilgartner and Brandt-Rauf, "Data Access, Ownership, and Control" (note 6), pp. 363–65.

11. See the discussion of the negotiations surrounding collaborations in Hilgartner and Brandt-Rauf, "Controlling Data and Resources" (note 6).

12. Hilgartner and Brandt-Rauf, "Data Access, Ownership, and Control" (note 6), p. 365.

13. See the discussion in Hilgartner and Brandt-Rauf, "Controlling Data and Resources" (note 6).

14. Human Genome Organization, "Guidelines for the Organisation of Single Chromosome Workshops," Version I, March 1992, photocopy, p. 2.

15. I attended a number of meetings where these issues were discussed, including one SCW in the United States and one in Europe, and a Chromosome Coordinating Meeting, at which data access policy for future SCWs was debated.

16. This is true in the case of maps based on STSs (discussed below) and of maps based on clones from publicly available libraries. On the latter see Hilgartner and Brandt-Rauf, "Controlling Data and Resources" (note 6).

17. Hilgartner and Brandt-Rauf, "Controlling Data and Resources," (note 6).

18. On data isolation see Hilgartner and Brandt-Rauf, "Controlling Data and Resources," (note 6).

19. National Institutes of Health and Department of Energy, *Understanding Our Genetic Inheritance: The U.S. Human Genome Project, the First Five Years, FY 1991–1995* (DOE/ER-0452P) (Washington, D.C.: Department of Health and Human Services and Department of Energy, 1990). A sequence-tagged site is a short DNA sequence that occurs only once in the genome under study. For more on STSs see Hilgartner, "Human Genome Project" (note 3).

20. Maynard Olson, Leroy Hood, Charles R. Cantor, and David Botstein, "A Common Language for Physical Mapping of the Human Genome," *Science* 245 (1989), 1434–35.

21. One estimate put the cost of operating the repository at $250 million over 30 years. See NRC, *Mapping and Sequencing the Human Genome* (note 4).

22. Olson et al., "A Common Language" (note 20), p. 1434.

23. For more on the complex reasons for this enthusiasm see Hilgartner, "Human Genome Project" (note 3).

24. Alessio Vassarotti and André Goffeau, "Sequencing the Yeast Genome: The European Effort," *Trends in Biotechnology* 10 (Jan.–Feb.1992), 15–18. For a discussion of organisms as "tools" see Kohler, *Lords of the Fly* (note 6). See also Adele E. Clarke and Joan H. Fujimura, eds., *The Right Tools for the Job: At Work in Twentieth-Century Life Sciences* (Princeton, N.J.: Princeton University Press, 1992).

25. For the network concept see Alessio Vassarotti, Bernard Dujon, Horst Feldman, Werner Mewes, and André Goffeau, "Structure and Organization of the European Yeast Genome Sequencing Network," photocopy, 1993; and Alessio Vassarotti, André Goffeau, Etienne Magnien, Bronwen Loder, and Paolo Fasella, "Genome Research Activities in the EC," *Biofutur* (Oct. 1990), pp. 1–4.

26. Vassarotti et al., "Genome Research Activities in the EC" (note 25).

27. The ownership policy, along with rules about quality control, was outlined in a ten-page internal document, signed by Bernard Dujon, Horst Feldman, André Goffeau, Ati Vassarotti, and Werner Mewes and titled "The Perfect Gentleman Sequencer," 1992.

28. One procedure for verifying sequence data was to have each sequencing lab anonymously resequence some regions previously sequenced by others. These spot checks could help assess the error rate overall and aid in identifying and correcting problems. All labs were required to do some "verification sequencing," for which they were paid but which conveyed no ownership rights.

Biotechnology's Private Parts (and Some Public Ones)

Thomas F. Gieryn

Is biotechnology private science, public science, or maybe a little of both? Who decides what is public or private about any science?

Perhaps responsibility for assigning the adjectives *public* or *private* to science should fall to the sociological analyst. The job seems straightforward: create two analytic spaces—"private science" and "public science"—give them defining qualities, take a look at biotechnology, decide where it fits best. Private biotechnology might be *secluded* science, intimate practices kept away from public view or intrusion. Many sites of animal experimentation are, for example, given the secured inaccessibility or invisibility ordinarily reserved for bank vaults. Private biotechnology might be *personal* science, the work of isolated individuals rather than a collective effort, each scientist alone with pipettes, ultracentrifuges, and PCR machines. Private biotechnology might be *owned* science, the property of some but not all, belonging to them and shaped by their private interests and values, and not necessarily pursued on behalf of the common interest. Although these analytic distinctions between public and private science rely on ordinary meanings of words like *secluded*, *individual*, or *self-interested*, the sociologist extracts them from everyday usage and imposes them on scientific practices as objective categories—precise, consistent, and logical. As an analytic tool, the public-private distinction is used *on* social life, though its origins *in* those practices are obscured.

There is, of course, another way to decide whether biotechnology is public or private science. It requires only that analysts extinguish their burning desire to settle the matter definitively. I prefer to work within a long-standing sociological tradition reaching back at least to the social phenomenologist Alfred Schutz, whose observation that the social world comes to analysts "preselected and preinterpreted . . . by a series of common sense constructs" implies that sociological "constructs [are] of the second degree, namely constructs of the constructs made by the actors on the social scene."[1] Sociologists confront a world already carved up into public and private parts. Analytic attention shifts to how those constructs are pragmatically deployed in everyday life. How do ordinary people use the distinction between public and private in accounting for science or for biotechnology? To answer that question, sociologists suspend a priori judgments about what public or private means and about the referents to which that distinction might reasonably be germane. They admit, in effect, that whether biotechnology is public or private science is "ultimately undecidable."[2] It is undecidable because actors use private-public in flexible, ambiguous, inconsistent, asymmetrical, and even contradictory ways—public is discursively reconfigured into private, as the situation changes. Nigel Gilbert and Michael Mulkay recommend that sociologists "become more sensitive to interpretative variability among participants and . . . understand why so many different versions of events can be produced."[3] The goal is not to improve upon disorderly usages of everyday language, as if the sociologist's mandate were to clarify what is really meant by private science: "Ordinary or lay language cannot be just dismissed as corrigible in the light of sociological neologisms, since lay language enters into the very constitution of social activity itself."[4] Instead, the task is to explore how one domain of social activity—the science of biotechnology—is constituted in and through messy and polysemic but practically potent constructs like public-private (and their synonyms). The methodological strategy is to "follow" *public* and *private* as indexical textual signifiers (remaining agnostic or indifferent about their referents) inserted by participants into their ongoing practices of making biotechnology.[5]

But *where* is the distinction between public and private connected to science and to biotechnology, not by analysts but by "actors on the social scene"?

Biotechnology in Its Place

The design of the new Biotechnology Building at Cornell University in Ithaca, New York, is one occasion where the distinction between public and private is discursively co-present with science.

Planning for the building began in 1983, shortly after the Biotechnology Program was organized at Cornell; ground was broken in 1986, and the building was first occupied in 1988. It has five levels with 98,000 usable square feet, sheathed in white with many windows of swimming-pool blue (Figure 1).[6] Its footprint is triangular with a sawtooth hypotenuse, and it forms two sides of a biology quadrangle. Most visitors enter into a two-story lobby featuring a prominent modernist doglegged staircase with vaguely nautical white-piped railings. The lobby, sometimes decorated with straw mats by Alexander Calder, is large and open, with natural light pouring in

Figure 1. Cornell University Biotechnology Building. Photo by the author.

through a glass wall. From the ground-floor lobby, one can enter a conference center (really a large auditorium), a smaller conference room or a vending area and lounge (for eating and, uniquely, for smoking). On the first-floor landing just off the lobby are administrative offices for the three units occupying the building: the Biotechnology Program and two sections of the Division of Biological Sciences—Genetics and Development; and Biochemistry, Molecular and Cell Biology. The Biotechnology Program's special research facilities are located on the ground and first levels. The top four floors are connected by an atrium stairwell, with a small sitting area, a secretaries' office, and a small seminar room adjacent on each floor. Faculty members' research labs and offices skirt the perimeter of the second through fourth floors, while the windowless interior space is used for heavy equipment, cold rooms, warm rooms, and dishwashing facilities. Davis, Brody and Associates of New York City were selected as the architects for the building, whose $34 million cost was shared by Cornell and the State of New York.

The spaces and places where science gets done have begun to attract the attention of historians and sociologists.[7] Most studies examine the effects of laboratory design: as iconographic expressions that reproduce dominant values; as spatial arrangements of social interactions that facilitate or retard the random exchange of scientific ideas and levels of creativity; as walls and doors that discriminate those allowed to participate in science from those who cannot. I take a different approach to the Cornell University Biotechnology Building (CUBB) by examining it in design, in the process of its making, or to extend Bruno Latour, as "laboratory-building in action."[8] Four aspects of the design process of science buildings are important for understanding the contingent distinctions between public and private.

First, design is not something accomplished by architects alone—that would confuse a professional jurisdiction with the actual practices involved in deciding what space is needed or how it should be arranged and outfitted.[9] At Cornell University, the Biotechnology Building was shaped by a design triad consisting of architects and engineers at Davis, Brody; end users, that is, faculty scientists and their various support staffs; and the Cornell central administration, ranging from a budgetary vice president to the campus architect. Design is collective work, vastly different from the common image of the "marquee architect" as solitary genius whose creativity is said to be the fount of wonderfully styled and functionally efficient spaces. Building design is more like what Howard Becker found in "art worlds." Works of art are made through the essential cooperation of a network of people (e.g., critics, paint suppliers and manufacturers, gallery entrepreneurs, the museum employee who removes a painting from storage and hangs it on the wall), though only one person usually gets credit as "the artist."[10] The meanings and purposes of public and private are negotiated among members of the design triad, who have diverse skills, discourses, and interests. Public-private is employed textually and graphically by architects, scientists, Cornell administrators—even the governor of New York—though in different ways and for different ends.

Second, design does not begin only when the architect begins to draft, nor does it end when construction commences or even when the building is occupied. The lament "we need some new digs" is as much a part of the design process as a CAD-rendered depiction of an intricate heating-ventilating-air conditioning system. Design is the *representation* of space, which assumes diverse embodiments, both prospective and retrospective: a rationale for a new building, a "conceptual program," an architect's phase report, brick-and-mortar, post-occupancy evaluations, and critical reviews. Design is a process in which spaces move along a gradient of stabilization, in both directions. The design is more malleable as talk or as words and pictures on paper or in a computer than when it is built into walls and doors, floors, and ceilings. But built designs never completely lose their interpretive flexibility. In its ongoing ordinary use and also in its manifest evaluation a building is forever becoming.[11]

Private-public is itself more or less stable as an element of the design of the CUBB. We shall watch it move first to the "material" (obdurate, resistant, black-boxed) and then back toward the "social" (negotiable, uncertain, pliable).

Third, design of a laboratory building is simultaneously the representation of physical space and human occupants, of the material and the social, of science and architecture. The design team is forced to shift from scientific to architectural registers (and back again), talking and drawing about the genetics of *Xenopus* and about the temperature or lighting systems necessary to keep the beasts alive. Science is rendered architecturally, just as the building is translated into anticipated scientific practices, with no privilege given consistently to one domain or the other. The design of science buildings is irreducibly heterogeneous.[12] Space is represented in a way that fits the needs of scientists, staff, and equipment; science and other elements of society are made to fit the equally imposing needs of the building. What is being designed? A building, but also a set of practices shaped to happen effectively within it, and even a society in which such situated practices also fit comfortably. The design of the building is equivalent to the design of who and what it will house and of the society in which it operates. Public-private is routinely used by designers to indicate distinctions among architectural spaces and at the same time among people, machines, and activities. There is ubiquitous interreferencing of the spatial (material) and the social.

Fourth, design is pragmatic and performative. Its paramount purpose is to bring into being a certain building. Design decisions are not determined by universal abstract principles of aesthetics or functional efficiencies or even cost. They are instead mediated by and through negotiated concerns for that representation most likely to get realized, with the highest odds of eventually assuming a stable and enduring existence. Design must satisfy the interests of those parties whose support is essential for the translation of abstract space on paper into a concrete place where science can happen. Evolving floor plans are representational devices enabling the enlistment of allies who, when hooked, have the means to move a project forward to obduracy: "Every blueprint can be read as another *Prince*: tell me your tolerances, your benchmarks, your calibrations, the patents you have evaded and the equations you have chosen, and I will tell you who you are afraid of, who you hope will come to your support, who you decided to avoid or to ignore, and who you wish to dominate."[13] "Publics" and "privates" are made up by designers to enroll possibly recalcitrant but needed allies. As a pragmatic and performative construct at work on the social scene, public-private should not be interpreted in terms of its referents but in terms of its role in moving a certain plan closer to materialization.

The methodological chore of following private-public through the design of the CUBB was made immeasurably easier by converting the pertinent documents into a digitalized form that was then inserted into a hypermedia computer environment. The raw materials included about 4,000 pages (6 linear feet) of documents produced in the course of design (e.g., correspondence, meeting minutes, progress reports, technical reports from consultants, and speeches at ceremonial occasions); several sets of blueprints and sketches; about 600 video images of the building in use; more than 40 transcripts of focused interviews conducted with scientists, architects, university administrators, and others connected to the design triad. An electronic search located the words *public* and *private* as they occurred anywhere in the data set, as well as *privacy*, *communal*, *shared*, and *collective*.[14]

This electronic exploration of where (and how) public-private was deployed by designers of the CUBB moves into two discrete strings of documents. That is, the distinction between public and private helps constitute a pair of design issues negotiated through a chained sequence of documents and, in the case of the interview transcripts, their latterly reconstructions during post-occupancy evaluations. First, "the public" is constructed as something that becomes simultaneously the beneficiary of and rationale for the building's existence *and* a risk or threat in need of containment or exclusion. "The public" is simultaneously *inside* the CUBB and kept *outside* a nested set of spaces assigned ascending degrees of "privacy." Second, representations of "communal spaces" ("common," "collective") for "shared equipment" figure prominently in the design process—before and after occupancy—as a site where testy issues of access, ownership, and control are played out among (mainly) faculty scientists seeking their "turf."

Public-private shapes the design of the CUBB as designers evoke and invoke the distinction (in diverse, asymmetric, vague, heterogeneous ways) to win arguments, seduce supporters, legitimate moves, control space, and anticipate the future.

"The" Public for and in Cornell Biotechnology

One explicit representation of public-private appears in a "bubble" diagram titled "Building Program," included in Appendix A of the "Program and Project Concept" phase report, prepared by Davis, Brody and Associates on 16 April 1985 (Figure 2).

For most large building projects, architects must provide their clients with periodic updates (phase reports) summarizing provisionally agreed upon decisions about the project as a whole. The "program" report is ordinarily followed by the "conceptual phase" report, although sometimes these two are combined (as at the CUBB).[15] After these reports are approved and changes are negotiated, the architects then prepare a "schematic phase report" (dated 16 July 1985), followed by a "design development phase report" (dated 14 November 1985). Then the architects prepare "construction documents," which provide the basis for "bidding the job." As a final step architects are often asked to provide the client with "as-built drawings," which show all of the "change orders" made during construction itself. This sequence of reports organizes the flow of a building project, informally and legally. They are important for marking time. The project calendar is organized in terms of when phase reports are expected from the architects and

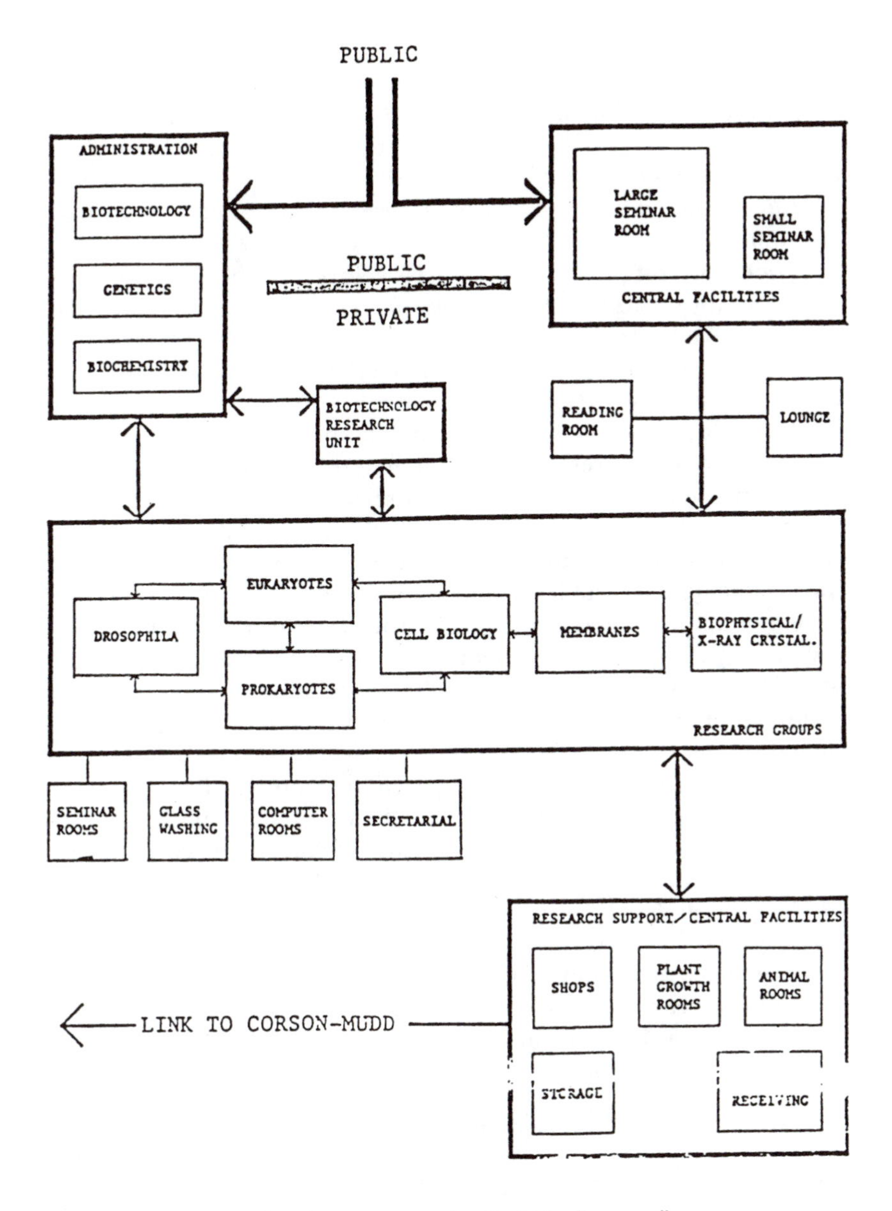

Figure 2. Program and project concept: the "bubble diagram."

when responses are due from the client-users. Each report is a provisional black-boxing of the project, gradually moving the plans toward more detailed specification and stabilization. Each incorporates (as it modifies) not just earlier phase reports but also the myriad informal agreements and piecemeal decisions made in the interim. All parties involved, but especially the architects, have a desire to "reach closure" as early as possible to avoid wasting time and money.

The appearance of public-private in Davis, Brody's conceptual phase report has a future and a past. How and why did the distinction ever enter into the design? What becomes of public-private as plans move toward construction — and then what happens after occupancy? The inscription in Figure 2 raises as many questions as it answers about the pertinence of public-private for biotechnology at Cornell. Who is the "public" and what is "private?" What do the arrows represent (access, functional adjacency, traffic flow)? What does the horizontal bar separating public from private signify? Is everything above the bar "public" and everything below it "private" — but what could this possibly mean?

The first mention of "public" in the seven-year paper trail occurs in a two-page memorandum (dated 21 May 1984) from the executive director of the Biotechnology Program (a nonscientist staff position) to a Cornell administrator. The memo identifies possible sites on the Ithaca campus for a new biotechnology building. After considering research and teaching as two elements in the selection of the best site, the memo goes on to talk about "public service" as a third factor: "A critical part of the Biotechnology Program is the opportunity for corporate scientists to be on campus and be involved with campus activities."[16] The memo equates "public" with "corporate scientists," and sets in motion an odd inversion: "public" refers to what is often called the "private sector" — that is, the corporate sector as opposed to the public sector of state government. In effect, public (as a constituency beyond Cornell scientists and a beneficiary of its proposed biotechnology building) becomes private (as in corporate). This inversion — a breach of logic maybe, but hardly ineffectual for getting this building off the ground — turned up in a post-occupancy interview with a Davis, Brody architect:

> They [Cornell] had a concept for the public function of the building . . . they wanted to have contact with similar organizations with similar work off-campus. Primarily private sector. There are a lot of companies in New York State who are interested in various aspects of high technology, they're doing it on a corporate scale. Cornell saw it as an opportunity to have shared appointments, shared research, grant money coming into Cornell, patents being generated.

This public-private inversion (or equivalence) proved to be strategic, as constituents and financial backers for the CUBB were enlisted from corporations such as Eastman Kodak and Corning Glass and from the State of New York.

Four months later, the public constituency for the CUBB is enlarged, as interest in the proposed building is distributed further beyond the Cornell scientists who would occupy it. In a planning document (dated 27 September 1984), Cornell students are added to private sector corporations in constituting a public for biotechnology: "Both undergraduate and graduate students are now demanding more exposure to this knowledge as part of their curriculum." But if the CUBB is *for* students, not all of them will have a place *in* it: "Undergraduate teaching functions are not programmed into this building." This early decision stuck. There are neither large classrooms nor teaching laboratories in the CUBB (although considerable "teaching" at all levels—postdoctoral, graduate, and undergraduate—goes on at lab benches and seminar rooms throughout the building).[17] A sociologically interesting pattern starts here and turns up repeatedly as the building process continues: the simultaneous expansion of publics with a putative stake in the CUBB *and* their exclusion from the building itself. At once, publics are multiplied outside the building as they disappear from within its walls.

Later in the same planning document, "public" is expanded again to include "senior industry, State and University officials." State officials are presumably acting on behalf of those who are really being referenced here: New York taxpayers and voters, who will in the end be asked to cover some of the $30 million budget. The building is for them, but only at a distance: "The public component must receive special consideration . . . due to the fact that biotechnology research incorporates the use of organic solvents, radioisotopes and other hazardous substances which preclude casual exposure. Research areas must be clearly defined and not positioned in such a way that visitors to the building face the potential of accidental exposure to dangerous substances." The taxpaying public of New York is not just a constituency and beneficiary of the CUBB, but a threat and a risk. The building is for the public, but it is not a public building in terms of accessibility and use. The design challenge is to shape a building that has sufficient space inside for those publics to see (and believe) that they are indeed part of it, but that also keeps those same publics out—away from research practices, materials and equipment that could bring harm to themselves or to science.[18]

The stage is set for the horizontal bar between public and private in Figure 2, but not yet in such a neat form. The simultaneous inclusion and exclusion of the public and the still-unclear definition of *private* require careful negotiation. Figure 3 is an early sketch (probably drawn by the executive director of the Biotechnology Program during Davis, Brody's presentation on 16 January 1985 of a preliminary conceptual phase report) and a lovely example of heterogeneous design. It is by now clear who constitutes the public and why they are an actant as important as "scrubbers" or "hoods." The building is for them, but not all of it is open to them. The two arrows in the figure route the public toward administrative space on the left and toward "central facilities" on the right (seminar rooms)—places that

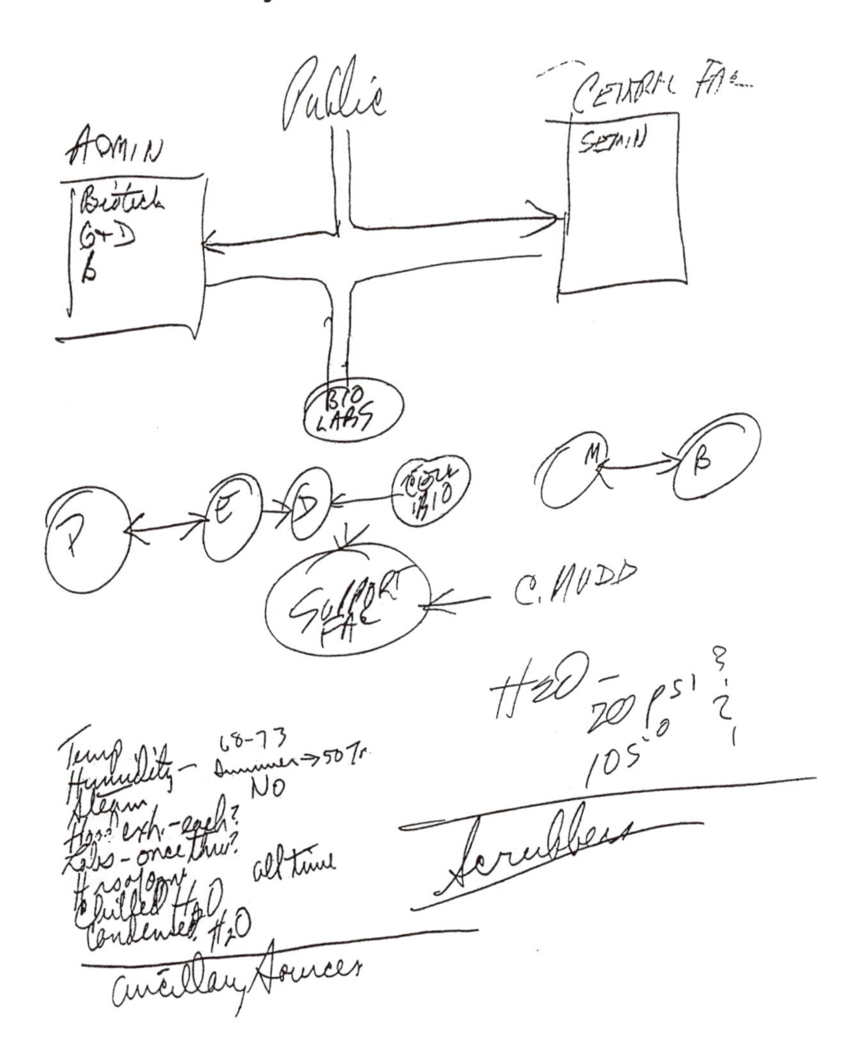

Figure 3. Early sketch showing "public" areas.

emerge as "trading zones" where the public comes into immediate contact with those below the bar (denizens of "bio labs," employees of "support facilities").[19] But, here at least, there is no direct path for the public to get from a starting point outside the building into the recesses of "steam" or "condensed H_2O" deep inside it.

To solve the problem of bringing the public inside the building while keeping them outside (most of it), designers pursue what I call the "beachhead strategy." Allow the public a toehold on the coastline, but not a step further into the interior. Create a place that allows and even invites students, representatives from industry, New York taxpayers, state officials to get inside—but cordon it off from private places where actual research goes on and where the public's presence is demonstrably unwelcome. Because of the promiscuous mix that makes up "the public," the beachhead turns out to be built on shifting sands; that is, the location of the bar between public and private spaces moves around depending on which public seeks access. Layers of accessibility are created, showing where some publics are allowed but other publics are not.

Several meetings of the building committee, leading up to the bubble diagram of 16 April 1985 (Figure 2), consider the issue of public accessibility. For example, under the heading "public safety" in the minutes for 30 January 1985, the Davis, Brody scribe notes: "Cornell will consider limited access to certain areas (card access). Should be clarified later"; "Avoid planning building to encourage traffic to other buildings as shortcut." In another preliminary draft of the conceptual phase report (dated 6 February 1985), the office of the director of the Biotechnology Program is described as "accessible to outside visitors," seminar rooms are "accessible to all occupants; public" (which makes the public into the residual "anybody other than a building resident"), a lounge and vending machine area is "accessible to students," but a stockroom for chemical stores and glassware is "accessible to faculty and researchers." The meeting record for 11 February 1985 notes that "public access from upper level athletic field walk (2nd floor) was desirable"—a recommendation that contradicted earlier wishes to prevent the building from becoming a cut-through for cold and snowy students.

In a later but still not final conceptual phase report (dated 21 February 1985), administrative spaces for the Divisions of Genetics and Development and of Biochemistry are labeled "accessible to faculty; students"; a P-3 containment lab is described as "accessible to all research groups and selected outside users." Finally, in a letter from two building committee members to the Cornell provost (dated 3 April 1985), the initial idea for a conference center (large seminar room) is endorsed as "an imaginative idea, and is just what we need for the program. It is the only large seminar room in the facility. Its location will allow a more public access so it can be used by both occupants of our building and our neighbors." In these discussions public varies from a cold student to someone in need of P-3 biohazard security, from occupants of nearby Cornell buildings to those paying a visit to the Biotechnology Program director. Decisions about accessibility are simultaneously social and architectural. Publics are defined by and through the spaces inside the building to which they have access.

The bubble diagram (Figure 2) "illustrates in general terms the desired functional relationships between the various groups . . . and the degree of public access." So significant is this diagram that it is lifted from Appendix A and reproduced (in miniature) in the main body of the report, suggesting that public-private demarcations fundamentally shaped the overall architectural design of the CUBB. The horizontal bar between public and private is still conceptual rather than architectural, functional rather than spatial, ideational rather than material. Its translation into something built-in is not straightforward: public and private on the bubble diagram do not correspond one-to-one to rooms in the building because, simply put, it depends on which *public* you mean. The arrangement of spaces inside the building depends in part on diverse publics who must variously be kept outside.

"The public" starts outside of the functional boxes that describe what will go on inside the building. Public entrances lead into a lobby, from which one should easily reach administrative offices and seminar rooms (the bubble diagram does not show people just passing through). Whether for a student checking availability of a class or for a faculty member from outside Biology coming to attend a lecture, these two spaces are public in a maximally inclusive way. From here access is more restricted, and the public shrinks drastically. Follow the arrow from the small and large seminar rooms down into the reading room and the lounge. The bubble diagram does not discriminate between the accessibility of these two areas, though the program description of the reading room says: "Locate on main circulation, near research labs, accessible to all occupants. Not a public facility." The lounge is described as a "space for faculty, staff and students to take breaks, lunch, meet informally. . . . Locate in a public area (near lobby) easily accessed from all parts of the building." Some public is welcome to eat in the lounge but not to browse the journal collection in the reading room. Some functions below the bar are completely private: "The research groups [drosophila, eukaryotes, etc.] should be located together on contiguous floors, preferably more remote from public access." But the "biotechnology research unit" (still below the bar) is more public, even though it presumably would pose the same risks to "casual" visitors as faculty research labs. The plant tissue culture room is part of the biotechnology research unit and "should be accessible to public areas." Which public? Not students, lecture audiences, or passers-through who might happen to be taxpayers—but consumers of those research services from public private biotechnology corporations outside Cornell. Moreover, this public gets its own dedicated space inside the CUBB: a "group of labs where industry users will have space allocated for their use." Corporate scientists are an outside public with a place inside the research areas of the building.

The beachhead shifts from lobby to administrative space and seminar rooms to lounges (not reading rooms) to biotechnology research service facilities to labs for corporate scientists: at each layer it sheds some of the publics that began the assault just outside the building's walls. The further inside, the smaller the public. The conceptual phase report summarizes the meaning of "public access": "The building functions are divided into two basic zones, the grade [ground] and first floors containing more public and support uses; and the second, third and fourth floors house the academic labs." The concept public-private is almost ready for rendering as architecture, first in floor plans, then in brick and mortar.

There are few changes worth noting as public-private moves through schematic and design development phases.[20] A better definition of the conference center comes in the meeting record of 21 June 1985: "The conference room remains attached to the lobby for good accessibility. The lobby provides space for pre-function activities, and includes direct access to public toilets, building lounge, kitchen and storage area. The lobby can be left open for evening events without allowing unauthorized access into the rest of the Biotechnology Facility." This stereoscopic image of the building—one eye public, the other private—is reinforced later in the same document: "The [conference] room will be used at times by groups other than Biotechnology, including evening events when the Biotechnology building will otherwise be closed." But the beachhead continues to shift, letting some publics further and further inside. A letter from the Cornell architect to the head of the Section on Genetics and Development (dated 2 December 1986) reads: "Although I feel the tile floor would present a handsome finish in the public spaces of the laboratory levels, I do not feel this should be done with disregard to functionality." This use of "public" would seem to violate the logic of "upstairs is to downstairs as private is to public" that had been more or less secured eighteen months earlier in the conceptual phase report. But in design logic is less vital than pragmatics. In this case the Cornell architect seeks to justify his choice of floor covering by inventing a very different sort of public: those who work in the most-private spaces (faculty research laboratories, offices, and equipment rooms) but who must leave those spaces via corridors going to the stairs and elevators and who, because of safety restrictions on eating in lab spaces, would want to take lunch in the atrium sitting areas. In the CUBB, public spaces (and the publics who are allowed in them) are consistently defined as outside some private space, although the latter telescopes from faculty labs and offices to all of the upper three floors to the building as a whole.

Construction is at last set to begin, and what better occasion than a groundbreaking ceremony to remind everyone that the public is inside the CUBB—and why. Governor Mario Cuomo of New York announces that "The Cornell University Biotechnology Institute will bring together—in one hundred thousand square feet of research space—scientists, farmers and scholars, members of the business community, academicians and students—all working together to make the promise of biotechnology a reality for the economy of our state."

This public for biotechnology, continues Cuomo, has both public and private parts: "It is a special pleasure for me to join in sealing this union with the participation of the third partner—the public sector—the State of New York. . . . This Institute will turn these future visions into practicality—and through technology transfer will create the private sector jobs our people

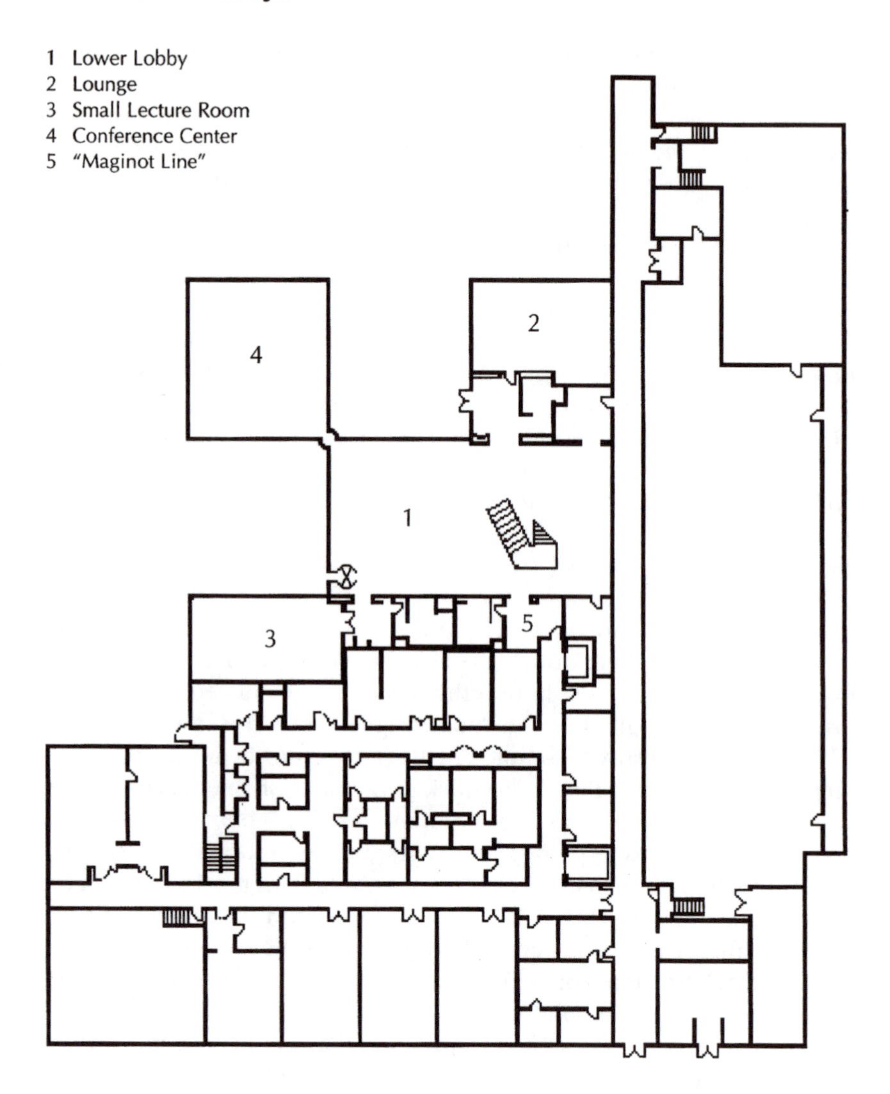

Figure 4. Ground floor plan.

need."[21] Why bother bringing up and bringing in all those publics, especially when they did nothing but create headaches for the designers who then had to work so hard at keeping them outside private places? Bruno Latour notes that "those who are really doing science are not all at the bench; on the contrary, there are people at the bench because many more are doing science elsewhere."[22] Governor Cuomo tells us who is doing science outside, so that there can *be* an inside to the CUBB: "Eastman Kodak,

General Foods and Union Carbide have pledged a six-year commitment of $2.5 million for the operation of the Institute, and Corning Glass Works has made a substantial contribution to its establishment. . . . And now, on behalf of the people of New York, let me sign this check for $32.5 million and present it to President Rhodes." The building would not exist if it were merely private, with no public interest—neither Cornell scientists nor the university itself have that kind of money. To get spanking new laboratories the Cornell faculty scientists must only surrender a beachhead, a place for their publics to get a toehold. With no beachhead some publics might suspect that the building was not really designed for them at all and would lose interest. But the beachhead cannot become a conquest, for that would require Cornell scientists to give up their desperately desired sequestered research space. The public becomes a constituency and beneficiary, then is treated as a threat and risk, so that it is allowed a beachhead but denied full-scale assault.

The spatial—and then material—inscription of public-private is now finalized (Figures 4, 5, 6). Rather than give my own account of the built place, I continue to rely on "actors on the social scene," starting with this gushing review in a background report (dated April 1989) prepared for the dedication ceremony: "The public spaces are magnificent, with an upper and lower lobby connected by an open staircase. Excellent facilities for conferences, seminars and workshops are present. . . . The architects have done an outstanding job in separating the public spaces from the more private research space." The head of the Division of Biological Sciences breaks down the floor plan into three levels of accessibility:

> There's sort of the public-privates. . . . Really three levels of privacy, therefore: the most public is down in the conference room, lobby, small seminar room, all that stuff, up to and open to . . . the offices. And then the second level was the biotech facilities, which are in this floor, which were still part of the inner part with research but since they are at least on the first floor, they could be more obviously get-toable. And then the top three floors, all of which . . . have a secretary where the stairs come up and the elevators so they could sort of see what's happening, and meant to be much more private.

The executive director of the Biotechnology Program constructs two Maginot Lines separating public from private:

> You'll be able to wander that area of the conference room and you can come up these stairs, but you can't get anywhere else in the building. The doors out here are locked that [would] allow you up in the labs, and the door downstairs is locked. So that we could allow anybody to use it and feel reasonably sure that the building wasn't going to be trashed, or we wouldn't have people wandering around in labs where they shouldn't be.

Figure 7 shows the first checkpoint: a lockable door on the first level, separating the upper lobby from corridors leading to the inner atrium, the

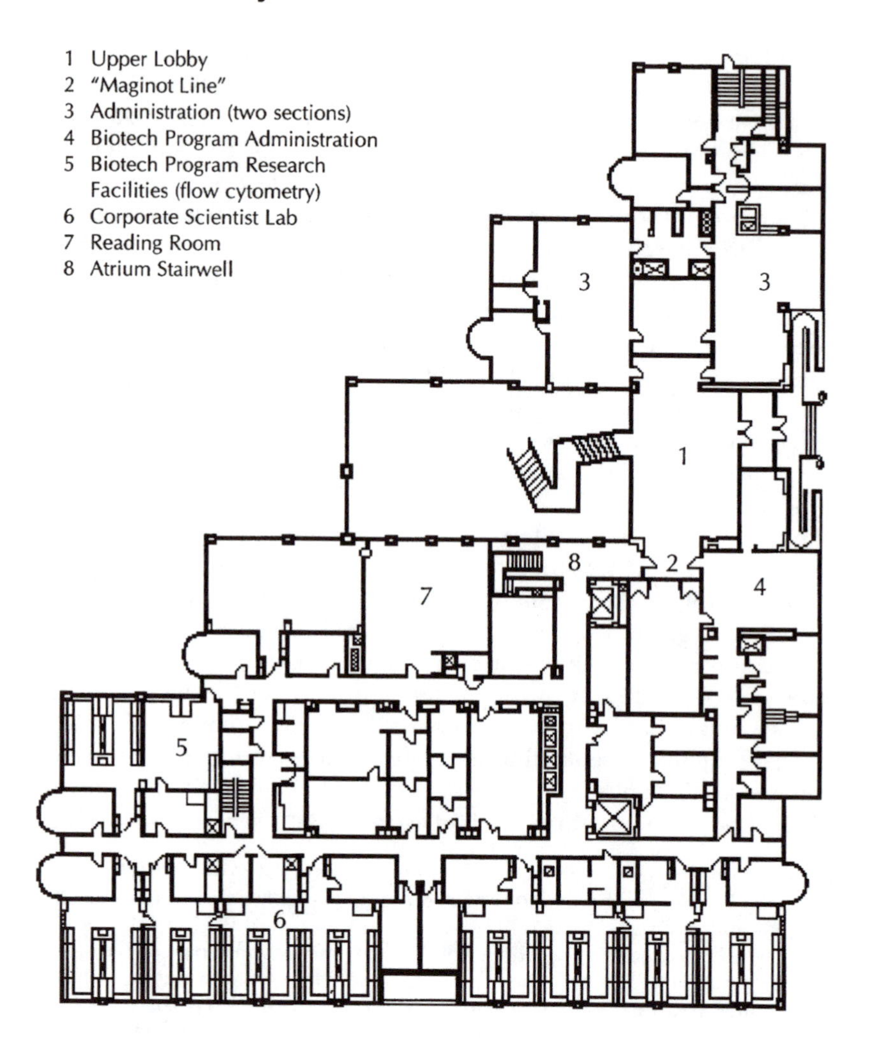

Figure 5. First floor plan.

reading room, research facilities of the Biotechnology Program, and via an immediately adjacent elevator the rest of the upstairs. Even when these doors are unlocked, access is guarded by the Biotechnology Program receptionist. The second checkpoint is off the lower lobby on ground level (Figure 8), which is also locked at night and during big public functions in the conference center. During the day this entry (which leads directly into research facilities) is protected by an absence of signage that would tell a casual public what is on the other side, and even more by a decor that makes one feel that the door leads to a basement-like place where only cognoscenti

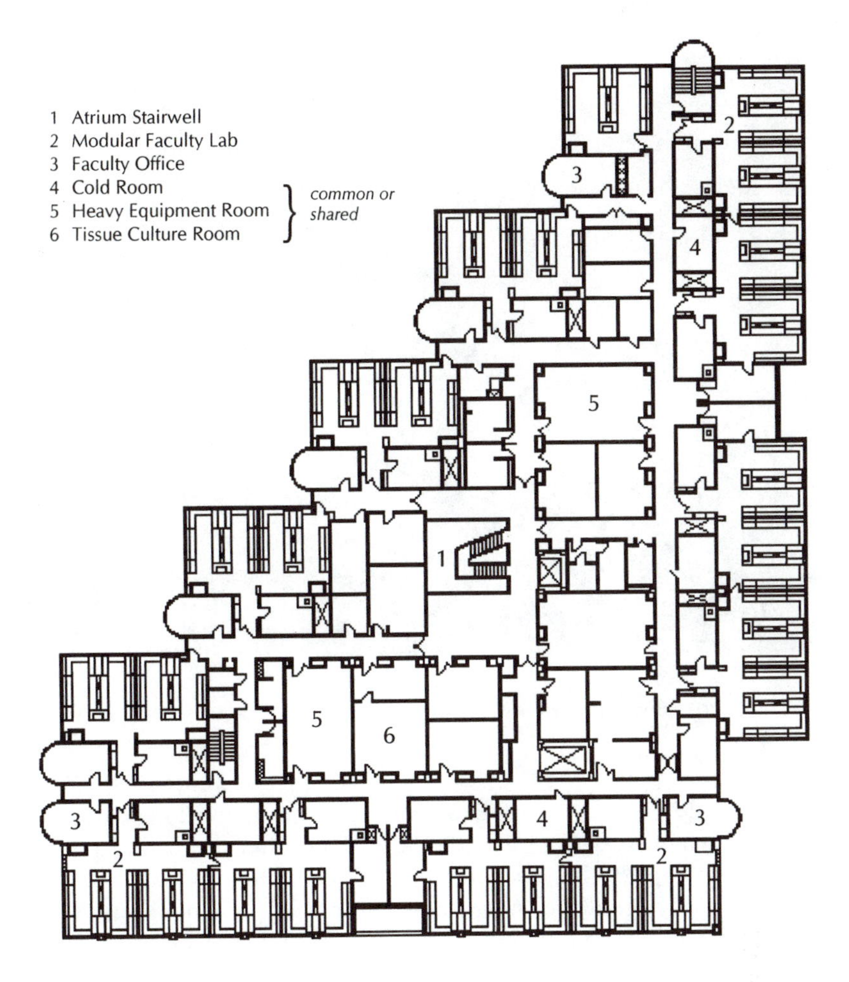

Figure 6. Third floor plan.

should be. The Calder straw mats, indoor plants, and upscale finishes of the (public) lobby contrast palpably with the painted cinder-block walls and ordinary floor coverings of the (private) research space behind the unmarked door.

Building occupants disagree on how well this separation of public-private is accomplished. One faculty scientist told me that he liked the way "the architect separated the public spaces from the private research functions. You . . . can have a party with five hundred people going on down there and have no impact whatsoever on the functioning [upstairs]." But another complained that "the research part of the building is poorly isolated . . . from the public part of the building. People ooz[e] through the building at

Figure 7. Maginot Line, upper lobby.

large. . . . People . . . take things like computers, no matter how hard you try." A third scientist told me that the lobby and other downstairs public spaces were too large: "It's a waste of money. It has no function."

The memories of some scientists, now content in their new laboratory building, are too short to remember why the lobby is there and why it is so large. Still another scientist commented that if the lobby space had been located in Wing Hall (an older Cornell science building) it would have contained five hundred centrifuges. But the lobby in CUBB cannot be for centrifuges, for it already houses the public. He continues: "And when I mean public space, I mean space which is used by people which have no connection with the building's normal function. And they use it a lot for entertaining."[23] No connection? How quickly he forgets.

Mine, Yours, Ours, Theirs, and Its: Colonizing Communal Space

"Private" is also juxtaposed to shared, common, or communal. Like the distinction between private-public, the private-shared distinction not only raises questions about access but also about ownership and control. Handwritten notes from 21 February 1984 first introduce this distinction into the

Figure 8. Maginot Line, lower lobby.

paper trail (Figure 9). Although the sketch bears no resemblance to what the building would become architecturally, its conceptual separation of personal laboratories and offices (for individual faculty scientists and their research staffs) from common space (not assigned to individual scientists) would last throughout the design process and would eventually be built in. Two months after this sketch was made, we learn what this common space is for. On a questionnaire sent to Cornell faculty biologists by the new Biotechnology Program (dated 12 April 1984), respondents were asked to list their needs for "shared equipment rooms" and "shared lab services." Specifically, they were asked to provide "design data" (footprint dimensions, weight, power, and plumbing needs) for "auxiliary research space" such as "dark rooms, incubation rooms, tissue culture rooms, etc." Auxiliary here implies separate and distant, and this question comes well after earlier ones about "individual laboratories" such as "what services and other facilities should be provided on laboratory benches and in hoods?" From the beginning, common rooms were spaces to be shared by several faculty scientists to house equipment or specialized research facilities not well located in private labs or offices.

As with the public beachhead, common spaces for shared equipment had

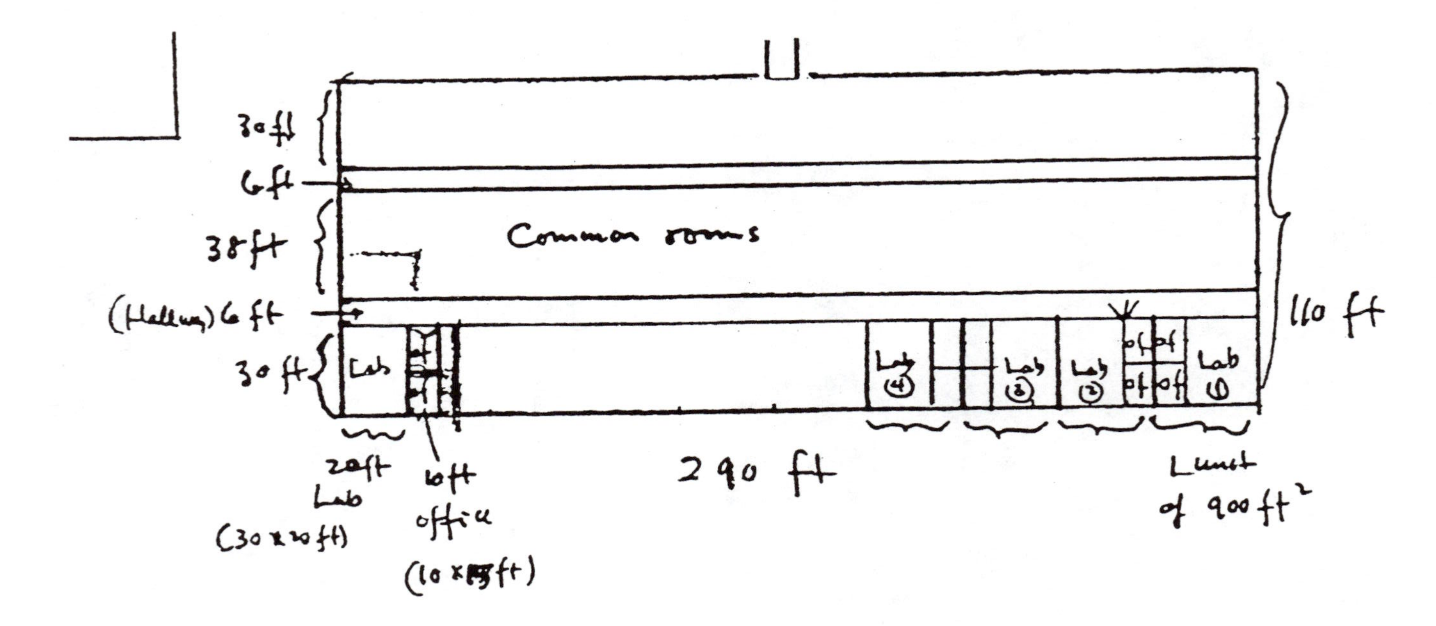

Figure 9. Early sketch showing "common rooms."

to do with the successful selling of the CUBB to those who required enrollment. Reflecting on efforts to create an attractive package, the Cornell provost told me:

> We had both departments [Sections of Biochemistry and of Genetics and Development] in less than ideal situations, and Biochemistry in particular was crowded to the point of not meeting program guidelines. . . . So we constructed the idea of the Biotechnology Program with those two departments at the core. Now, if the State was interested in providing facilities that supported this kind of thing, we should make use of this opportunity both to solve the local problem and . . . to provide the kind of facilities that needed to be developed as shared resources, things like the sequinator.

The CUBB was justified in part by the obvious advantages of gathering in one spot faculty scientists and their expensive experimental equipment (which had previously been scattered all around campus). By then making these research facilities and equipment available to "researchers from the corporations" in biotechnology, Cornell also had New York State and its publics interested in "common spaces."

But to see common spaces merely as useful for enlisting vital allies misses much of their significance in the design process. From the start, laboratory space assigned to individual faculty scientists was treated as identical—in square footage, configuration of benchwork and free wall space, and provided services and utilities. After Davis, Brody came aboard, laboratory spaces were described as "modular," "standardized," and "generic," as if all thirty-five of them came from a cookie-cutter. The modular lab idea took hold because it seemed to satisfy the interests of all three corners of the design triad: the architects (unique specifications for each lab were not needed); the administrators (modularity would cut construction costs); and the faculty scientists (modularity meant parity). Laboratory spaces are more than places for doing science; they are also measures of rank in the local hierarchy. Modular labs were an attempt to nip "turf wars" in the bud and a means to prevent invidious spatial distinctions among Cornell faculty scientists.[24]

Unfortunately, the modular lab decision did not preempt struggles for more room but merely displaced turf wars onto the common spaces. The promise of additional communal space outside the private lab for heavy equipment and specialized research facilities was a means to cool out scientists who might resent being denied the chance to customize their lab space, and who (in some cases) were given less personal lab space in the new CUBB than they had had before.[25] Scientists denied the chance to shape their private space to fit the particularities of their local lab culture and ongoing research were instead given the opportunity to tailor selected common spaces to the peculiar instrumentational demands of their science.[26] The availability of additional tailored space outside private labs may have mollified some Cornell faculty scientists, but raised its own troubling issues.

Who would share how much communal space with whom and for what purposes? The struggle for more space did not cease; only the means to pursue it changed. You could not argue as an individual scientist that you needed more bench space for five postdocs or a special room for a personal-use laser. Instead, you could get more space outside the private lab only *through other scientists* whose research and facilities needs were sufficiently similar to yours (and sufficiently different from others) to warrant a collective claim on communal spaces equipped to meet those particular needs.

Thus, a crucial moment in the building's design was the formation of the six research groups, each of which would be assigned common spaces customized for their shared use. The written record of the design process (and subsequent interviews) suggests that assignment of the thirty-five faculty scientists to these six groups—prokaryotes, eukaryotes, drosophila and development, cell biology, membranes, and biophysics—was accomplished, perhaps surprisingly, without rancor. Though some minor jockeying occurred (one slot was moved from cell biology to biophysics, another from prokaryotes to drosophila and development), the constitution of the six groups was settled early on (in April 1985) by clustering scientists in terms of the big, expensive, and (from a utilities standpoint) demanding machines they needed for research. One faculty scientist told me that the formation of research groups was based on

> shared equipment, both equipment and space. It makes sense to put centrifuges together in a room, but not computers and centrifuges maybe. And so biophysics types would have different shared needs than the molecular biologists, for example. . . . There wasn't any difficulty that I encountered. There were some shifts because people said: "Well, I'm really doing this, or I'm going to be doing that." . . . The extremes were obvious. The biophysical group is the other end from the amphibian development group. And the people in the middle were located physically in the middle.

Biophysics ended up on level two, drosophila and development was on level four, and the remaining groups were on level three with spillover up and down.

Formation of these groups shifted the "space bargaining unit" from the individual faculty scientist to the research group seeking more communal space. I call their strategy "colonizing the communal": relocate as much of your stuff as you can from the private lab to the common rooms and other shared spaces to free up square footage in your lab for activities and equipment that simply must take place there. In effect, the more stuff you can distribute throughout the building, the more space comes under your control and use—though not your private ownership, for it remains communal and shared. The six research groups used several arguments to establish and expand communal colonies outside the private labs. A given machine had to be exported because it did not fit into the modular shape of the private lab, because it was too noisy and dirty, because it required unusual utilities hook-ups, because it was used only sporadically, or because it was used by more than one faculty scientist. These arguments were variously successful. Although the building committee sought to remain true to its egalitarian impulses by initially assigning communal space with a universalistic formula—"number of scientists in the group × ~600 square feet"—in the end, some groups received more communal space per capita than others.

Although the CUBB contained two other types of space outside the private labs that could be colonized—rooms on the ground and first floors housing specialized equipment that require tending by expert technicians (e.g., flow cytometry facility), and general support units on the upper three floors (e.g., dishwashing facilities)—I concentrate on the "communal research spaces" that take up most of the windowless interiors on the second, third, and fourth floors. Positioning for communal space began even before the six research groups were formed. On 1 May 1984, just two months after the first rough sketch showing "common rooms" (Figure 9), a memo from the head of the Division of Genetics and Development to the director of the Biotechnology Program outlined the division's anticipated needs for common space (recall that two divisions would move into the CUBB). It proposed that each faculty member get 1,650 square feet, but their dishwashing, cold, and incubator rooms would be outside that assigned space. However, the memo continued, "extra space [is] made available by taxing each faculty unit by 150 square feet" to be used for cesium source, film darkroom, photography darkroom, and four heavy equipment rooms. The memo does not detail why such activities are not well suited for private lab space (cesium can be dangerous, darkrooms need light-omitting enclosure, heavy equipment is loud and dirty); that is understood by all. About two weeks later (but still well before Davis, Brody architects were hired), the Cornell scientists assembled for the first time (in "Conclusions of 5/16/84 Building Committee Meeting") a list of "Building Groups." This document would be repeatedly emended over the next six months. Faculty members of the six groups are identified, but in the typed meeting minutes no mention is made of communal or shared spaces. However, handwritten notes evidently added later list special equipment or facilities beside each research group. For example, prokaryotes need "storage 300 feet, one warm room, one seminar," while the developmental group (not yet merged with drosophila) needs "one cold room, one warm room and one cool room."

The 12 July 1984 meeting minutes show the research space segmented by the six groups; under each is a list of faculty scientists, including several unnamed "new faculty," with the designation that each is to receive a lab of 1,100 square feet and an office of 150 square feet. Also under each research group is a category appearing for the first time in these accountings: "*Communal." The asterisk points to the note: "Instrument rooms, heavy equipment, etc., to be used by more than a single research group." Though

TABLE 1 Space Allocations in Common Rooms by Research Group

Research group	*12 July 1984, Cornell* Number of faculty	Square feet per faculty member	Total square feet	*21 February 1985, Davis, Brody, pre-program* Number of faculty	Square feet per faculty member	Total square feet	*16 April 1985, Conceptual phase* Number of faculty	Square feet per faculty member	Total square feet
Prokaryotes	6	625	3,750	6	704	4,225	5	561	2,805
Eukaryotes	7	636	4,450	7	751	5,255	7	576	4,030
Drosophila development	7	657	4,600	7	752	5,265	8	560	4,475
Cell Biology	6	625	3,750	5	780	3,900	5	590	2,950
Membrances	4	625	2,500	4	800	3,200	4	525	2,100
Biophysics	5	805	4,025	6	853	5,120	6	672	4,030

the space is allocated to a research group and its square footage is a function of the number of faculty in the group (550 square feet per scientist), it is evidently to be shared by a more inclusive group of building occupants. This early ambiguity will haunt the communal spaces even after occupancy as users struggle to understand what is theirs, ours, yours, mine. Table 1 shows how communal space is assigned to each research group at this meeting and at several subsequent key moments. Two patterns are interesting to watch: the rise and fall of total communal space in the building, and changes in the shares of communal space for each research group.

Apparently, at the 12 July meeting word went out from the building committee to each research group: tell us what you intend to fit into the communal space. Clearly, this request was an opportunity for colonizing as much of the communal space as possible, and none of the groups were shy about indicating a crying need for more. The drosophila and development group responded first, with a 23 July 1984 memo assigning specialized equipment to the "proposed 3,850 square feet" of common rooms for ultraviolet work, autoradiography, centrifuges, freezers, incubators, shakers, sectioning, balances, electrophoresis (sequencing),[27] and computers. The memo then becomes both creative and aggressive, seeking to move equipment and facilities from the private labs into space that would not count against the allotted 3,850 square feet. The group suggests that the fly medium prep, dishwashing rooms, cesium source, one constant temperature room, and fly rooms "should come from the common space designated for the building in general." The rationale is that other research groups are likely to get space for their special facilities too: "As you know, foodmaking, dishwashing and bottle storage facility is an essential part of any drosophila operation, just like growth chambers and animal rooms are to plant and mammalian geneticists

21 June 1985			*15 August 1985, Schematics*			*10 October 1985, design-development*		
Number of faculty	*Square feet per faculty member*	*Total square feet*	*Number of faculty*	*Square feet per faculty member*	*Total square feet*	*Number of faculty*	*Square feet per faculty member*	*Total square feet*
5	555	2,773	5	450	2,252	5	381	1,905
7	580	4,062	7	533	3,729	7	608	4,255
8	528	4,178	8	514	4,109	8	503	4,025
5	567	2,836	5	564	2,822	5	535	2,675
4	527	2,109	4	538	2,151	4	530	2,120
6	668	4,007	6	663	3,980	6	672	4,030

respectively." Another rationale for locating equipment in non-group (building-wide) communal space is that scientists from outside the group will also need it: "The cesium source would be used by many people in the building and we feel it should not come from our communal space." Allies are enlisted from nature as well: "We felt it best that each lab has its own separate fly lab—to reduce the likelihood of a 'global' mite epidemic." The memo ends aggressively, pushing for more communal territory: "The additional space needed for the fly rooms may have to come from the 1,100 square feet allotted for biochemistry. This, we believe, further strengthens our contention that rooms 2A–C [cesium source, etc.] should come from the general common space allotment."

Other research groups made similar kinds of pleas. For instance, the cell biology group argued that "a 37° room should be included in the Biotechnology [Program] Tissue Culture Facility." But none were as elaborately justified as the memo from the fly people. Did it work? In this first reckoning (Table 1) the communal space per faculty scientist puts drosophila and development in second place only to the biophysicists (who stay on top all the way through); however, by the final accounting just before design development (4 October 1985), they had slipped to fifth place.

Architects from Davis, Brody were hired in December 1984, and in early 1985 they sought to refine the earlier descriptions of the functions of common rooms. For example, "heavy equipment rooms" were described as "communal group laboratory space for housing shared equipment and/or equipment which need not be kept within the assigned faculty lab modules." The basic distinction of private (faculty scientists' labs and offices) and shared space (for the research groups, and maybe others) is maintained. Meeting minutes of the building committee for 19 January 1985 seek to clear up possible confusion about the proper use of common rooms: "Communal space assigned to each group must meet shared facility needs for all members of the group. These spaces are to be used for activities common to the total group and not as an extension of assigned individual faculty laboratory space. Such activities as thin sectioning must be included in individual faculty member's assigned space." Here the building committee is trying to prevent intragroup turf wars: shared equipment space is ours, not mine and yours. But it also seeks to contain colonization of the communal by indicating procedures that may and may not be farmed out. Glass-washing rooms are here described as "central," that is, included in building-wide space rather than space allocated to each research group. In contrast, each group is told to make room for a 120-square-foot cell culture transfer area in their assigned communal space. This kind of back and forth continues for the next year or so, with research groups justifying the need for more communal square footage as they dump equipment and activities into "central" building-wide spaces. Meanwhile the building committee tries to keep the overall building size within budget by forcing equipment and functions back into the private labs and group communal spaces. These are the "space wars" at Cornell Biotech. Handwritten notes from a building committee meeting of 24 January 1985 sound ominous: "dynamics of sharing vs. identity."

As design moves toward the program and conceptual phase report of 16 April 1985, the combined communal spaces claimed by the six research groups balloon, since wish lists are not yet tempered by budgetary realities. Biophysicists win some and lose some. One member of this group is told that an environmental chamber for radioisotope research would "have to be located in Dr. ——'s research space allocation."[28] But the group as a whole picks up space by having a "high level instrument testing and repair shop" moved downstairs to the Biotechnology Program space, with the 275 square feet still made available to them as a shared instrument room. Davis, Brody descriptions of common rooms note why their functions cannot be housed in private faculty laboratories: "Communal laboratory space for housing shared high quality instruments and/or instruments which must be operated under a controlled environment"; "area is hazardous due to high qualities of electricity produced"; "mammalian and plant cell work must be carried out in separate facilities"; "space must be kept clean, uncontaminated and positive pressure to be considered"; "shared facility for experimentation involving levels of radioactivity higher than those permitted in standard labs." Each research group gave the architects descriptions of their needs, interesting sociologically not for their accuracy but for their pragmatic utility in efforts to colonize the communal. Budgetary constraints and the inelasticity of the building's total square footage eventually intrude: "The autoclave/media prep, dishwashing and dark rooms should be assigned at a rate of one per floor. This will mean consolidating the communal space reserved for these purposes when more than one research group is located on the same floor." Now these facilities are shared by everyone on the floor, not just members of a single research group. By the conceptual phase report, communal research space has been reduced dramatically—24 percent is gone.

From the program through schematic and design development phases, the communal space assigned to the six groups collectively declines but does so unevenly. Prokaryotes loses big time (32% lost), drosophila and development and cell biology lose much less, eukaryotes and membranes gain a little, and biophysics remains unchanged at the top. On 22 August 1985, prokaryotes lost space for half of an overflow standard lab module, half of a heavy equipment room (given to eukaryotes), and a film-processing room (they were instead to share one with drosophila and development). Later they lost one of two transilluminator rooms and the special space for electrophoresis. The paper trail provides little evidence of why these spaces were lost or reassigned. Meeting minutes state only that the changes are "agreed to" by the users group.

The process takes a significant turn with a memo of 12 December 1985 from the building committee to representatives from each of the groups, asking them to provide complete lists of machines slated for shared equipment rooms—including the name of the scientist who "owns" each piece. Before this point equipment lists were collective for each group. The new memo elicited a revised set of drawings such as the one in Figure 10. Ambiguity about ownership arises again. The heavy equipment room belongs to the research group, but space inside it is assigned to pieces of equipment owned and tagged-as-theirs by individual faculty scientists. The *space* is shared, but is the *equipment*? Realizing the potential for an all-out turf war even before construction begins, the building committee tries to gloss over the problem in its meeting minutes of 3–6 February 1986: "The faculty committee recognizes the impossibility and inadvisability of custom designing each [heavy equipment] room to its proposed equipment, two years in advance of occupancy. The rooms will be designed generically." So do not take too seriously those precise locations of "my centrifuge" and "your freezer." Fat chance! On 15 May 1986, the eukaryotes group informed the executive director of the Biotechnology Program that "this equipment room is already too crowded. The architects were not even able to put in all of the equipment in the drawing. —— tells me that all of ——'s equipment was supposed to go in the adjacent empty room." The dilemma is obvious. To identify each piece of equipment by the name of its owner raises questions about the distribution of "communal" space within the research group and invites a space war; to fail to do so may result in too little space in the shared rooms—naming names becomes a means to justify requests for still more common space.

Fast forward now to the completed and occupied building. Even within a

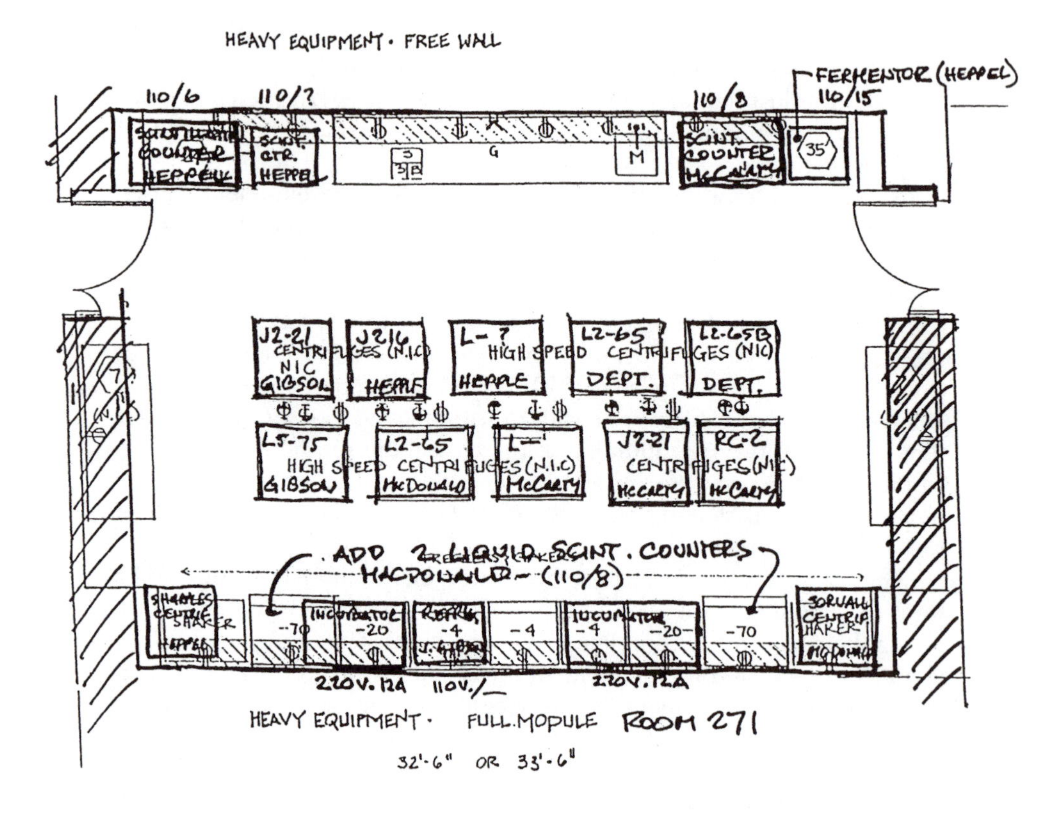

Figure 10. Communal room showing space dedicated to individual scientists.

faculty scientist's laboratory there are demarcations of shared and private—simply because of the dozen or so students, postdocs, and technicians who work there (Figure 11). Pointing to a benchtop to the left of a masking tape line, a postdoc told me, "That is public space; other people are using the centrifuges, bunsen burners." To the right is her space for her tools. But the lab as a whole is a private place, shared by many but locked off from corridors, for several reasons. As a Davis, Brody architect pointed out: "I think it's out of a concern for privacy, out of concern for contamination, out of just a little bit of secrecy. I mean, you would know more than I do about what's involved in getting grants. . . . Each faculty researcher gets a boundary, to protect his interests, what he has at stake."

What finally happens to the common rooms for shared equipment? For the most part the Cornell scientists are happy to have them, as they seem to work according to plan. A member of the biophysics group told me that "it's nicer to have the centrifuge and noisy equipment in a big central room that's on this floor . . . better than having it right in the lab." Someone from

Figure 11. Masking tape demarcation between public (right) and private (left).

eukaryotes said: "Now here is a perfect example of sharing because —— [a student] is in ——'s lab, but that's my hood." And a prokaryote researcher admits that "I haven't discerned any territoriality. . . . We have a little more shared equipment space than we had [before]." There were some problems, however, when actual equipment was first moved into the CUBB, and space for yours, mine, ours, theirs, and its suddenly lost its abstract character. When asked about "troubles," the building manager (who heads the custodial staff and is responsible for keeping equipment in working order) told me, "particularly when people move in . . . you needed to figure out, prior to bringing the stuff, where it was going to be set. This guy might have more power than this other guy, and he's got a whole wall . . . and I've got this corner." Parity is easier to maintain on paper. One scientist said, "It wasn't ever really spelled out exactly either, so people sort of started moving things into rooms, assuming this was where they were going to go, and it wasn't all that clear."

Moving in was just the start of troubles with common space. A cell biologist noted that his communal spaces were farther from his lab than from other labs in his research group: "There's one odd thing about this building, which is the distribution of all the common spaces, the common equipment rooms and the cold rooms and the warm rooms. Because the building is not a rectangle, there isn't actually equal access to all that stuff. It turns out that the labs on the diagonal [hypotenuse] tend to be kind of farther away from things." This problem is a trifle compared with what has become a "tragedy of the commons." As another scientist noted, "the lyophilizer had been used repeatedly by someone who, well, almost wrecked the pump by pumping organic solvents into it. So people actually do this, when there's a common room, and it's not their own labs, they'll abuse things in that room. And, so I feel that doesn't work so well." An extreme view comes from a biophysicist (who once asked *me*, "What would happen to our universities if they closed down all the sociology departments?"): "I simply don't believe in shared facilities. . . . Somebody uses a centrifuge and they spill something in the rotor. It's not their centrifuge. Do they clean it up? Not on your life. They walk away and leave it, and the next guy finds that the lining has been attacked by something or other or else there is powdered glass in the bottom. That sort of thing goes on all the time."

Perhaps because of such unpleasant experiences the definition of *common* has become more restrictive at the CUBB. One scientist told me that certain common rooms on her floor had been rekeyed to reduce user access. In biophysics, common rooms have disappeared altogether: "We didn't really want so many truly common rooms. We each wanted another lab. I wanted one that I could keep dark, and someone else wanted one for a laser, someone else for x-ray diffraction equipment. So we're all in complete agreement that this is a fine way to divide it up. So these are not common rooms at all." It may be instructive that the word *communal*—used throughout the design process before construction—appears not at all in transcripts of interviews done after occupancy.

Conclusion

How far upstream should sociologists and historians take the beliefs and practices of scientists? At least, I suggest, to the design of places where science gets done. Knowledge-making projects are situated not just by the gender, race, class, and nationality of the knower, but also by the material and architectural surroundings of inquiry. All knowledge begins as local knowledge. The questions asked, and the answers permitted, are shaped by the machines and people specifically gathered *there*. The design of new science buildings is an archaeological site for examining struggles over the definition of science: its audiences, purposes, beneficiaries, and culture. Among other places, it is where constructs of public and private science are deployed, battled over, and resolved.

If you go to Ithaca and visit the Cornell University Biotechnology Building, you will not see the words *public* or *private*, *shared* or *communal* on signs anywhere. Yet so thoroughly is public-private inscribed in the building that—looking on it as it emerged from years of design—it hardly makes sense without the distinction. Almost invisible now, this distinction between public and private is everywhere etched into walls, doors, windows, rooms, instruments, scientists, and science.

This article is based on a report from the Project on Laboratory Design in Biotechnology, which is housed at the Department of Sociology, Indiana University, Bloomington, and funded by the Andrew W. Mellon Foundation. Information on primary sources is found in note 14.

Notes

1. Alfred Schutz, *Collected Papers*, 3 vols., vol. 1: *The Problem of Social Reality* (The Hague: Martinus Nijhoff, 1971), p. 6.

2. Michael Lynch, *Scientific Practice and Ordinary Action* (Cambridge: Cambridge University Press, 1993), p. 115.

3. G. Nigel Gilbert and Michael Mulkay, *Opening Pandora's Box: A Sociological Analysis of Scientists' Discourse* (Cambridge: Cambridge University Press, 1984), p. 2.

4. Anthony Giddens, *Central Problems in Social Theory* (Berkeley: University of California Press, 1979), p. 246. I have written a series of articles on the episodic and pragmatic demarcation between science and nonscience, beginning with Thomas F. Gieryn, "Boundary-Work and the Demarcation of Science from Non-Science: Strains and Interests in Professional Interests of Scientists," *American Sociological Review* 48 (1983), 781–95; and synthesized in Gieryn, "Boundaries of Science," in *Handbook of Science and Technology Studies,* ed. Sheila Jasanoff et al. (Thousand Oaks, Calif.: Sage, 1995), pp. 393–443.

5. The strategy is akin to semiotic actant-network studies such as Michel Callon, "Society in the Making: The Study of Technology as a Tool for Sociological Analysis," in *The Social Construction of Technological Systems,* ed. Wiebe E. Bijker, Thomas P. Hughes, and Trevor Pinch (Cambridge, Mass.: MIT Press, 1987), pp. 83–103; and Bruno Latour, *Science in Action: How to Follow Scientists and Engineers Through Society* (Cambridge, Mass.: Harvard University Press, 1987).

6. The following description captures the building as I observed and videotaped it on repeated visits to Ithaca, starting with the dedication ceremony in 1989 and most recently in 1992.

7. A pioneer in historical studies of laboratory design is P. Thomas Carroll; see most recently his "Room to Maneuver: Laboratory Design and Research Agendas in American Academic Chemistry, 1857–1941," presented at the conference "Making Space: Territorial Themes in the History of Science," University of Kent at Canterbury, March 1994. See also Tom Allen, *Managing the Flow of Technology* (Cambridge, Mass.: MIT Press, 1977); Larry Owens, "Pure and Sound Government: Laboratories, Gymnasia, and Playing Fields in Nineteenth-Century America," *Isis* 76 (1985), 182–94; Owen Hannaway, "Laboratory Design and the Aim of Science: Andreas Libavius Versus Tycho Brahe," *Isis* 77 (1986), 585–610; Alejandra C. Laszlo, " 'Physiology of the Future': Institutional Styles at Columbia and Harvard," in *Physiology in the American Context, 1850–1940,* ed. Gerald L. Geison (Bethesda, Md.: American Physiological Society, 1987), pp. 67–96; Steven Shapin, "The House of Experiment in Seventeenth-Century England," *Isis* 79 (1988), 373–404; Sophie Forgan, "The Architecture of

Science and the Idea of a University," *Studies in the History and Philosophy of Science* 20 (1989), 405–34; Arthur Stinchcombe, *Information and Organizations* (Berkeley: University of California Press, 1990), pp. 320–40; Joel Shackleford, "Tycho Brahe, Laboratory Design, and the Aim of Science," *Isis* 84 (1993), 211–30; and the following articles in *Science in Context* 4 (1991): Bill Hillier and Alan Penn, "Visible Colleges: Structure and Randomness in the Place of Knowledge," pp. 23–50; Michael Lynch, "Laboratory Space and the Technological Complex: An Investigation of Topical Contextures," pp. 51–78; Adi Ophir and Steven Shapin, "The Place of Knowledge: A Methodological Survey," pp. 3–22; and Steven Shapin, " 'The Mind Is Its Own Place': Science and Solitude in Seventeenth-Century England," pp. 191–218.

8. Latour, *Science in Action* (note 5). Among the rare studies of the design process for science buildings are Ellen Shoshkes, *The Design Process: Case Studies in Project Development* (New York: Whitney Library of Design, 1989), pp. 96–123, on the Lewis Thomas Molecular Biology Laboratory at Princeton; and—of the coffee-table variety—Elizabeth L. Watson, *Houses for Science: A Pictorial History of Cold Spring Harbor Laboratory* (Plainview, N.Y.: Cold Spring Harbor Laboratory Press, 1991).

9. On the professionalization of tasks, see Andrew Abbott, *The System of Professions: An Essay on the Division of Expert Labor* (Chicago: University of Chicago Press, 1988), p. 50, for the observation that "the architect becomes a broker negotiating a general design through a maze dictated by others." See also Robert Gutman, *Architectural Practice* (New York: Princeton Architectural Press, 1988); and Dana Cuff, *Architecture: The Story of Practice* (Boston: MIT Press, 1991).

10. Howard S. Becker, *Art Worlds* (Berkeley: University of California Press, 1982).

11. Lynch, "Laboratory Space" (note 7). Michael Lynch sees lab places as incessantly "made": "This does not make floors and walls irrelevant, but it suggests that when various practitioners construct and tear down walls, assign equipment to places and practitioners to equipment, manage computer files, set up precautions against contamination, and so forth, they are . . . effecting local arrangements within a specific topical field. . . . By trying to understand the space of knowledge we confront an ecology of local spaces integrated with disciplinary practices." The idea that design moves along a gradient of stabilization comes from constructivist (often Callonian or Latourian) studies of technology (in effect, I treat buildings as walk-through machines); see, e.g., Bijker et al., *Social Construction* (note 5); and Wiebe E. Bijker and John Law, eds., *Shaping Technology/Building Society* (Cambridge, Mass.: MIT Press, 1992).

12. See John Law, "Technology and Heterogeneous Engineering: The Case of Portuguese Expansion" in *Social Construction*, ed. Bijker et al. (note 5), pp. 111–34. As Law describes it: "In explanations of technological change the social should not be privileged. . . . Other factors—natural, economic, or technical—may be more obdurate than the social and may resist the best efforts of the system builder to reshape them. . . . The stability and form of artifacts should be seen as a function of the interaction of heterogeneous elements as these are shaped and assimilated into a network." Quotation on p. 113.

13. Bruno Latour, *The Pasteurization of France* (Cambridge, Mass.: Harvard University Press, 1988), pp. 200–201. My discussion in this paragraph refers to prospective design—the representation of space before its material construction.

14. These primary materials have been archived at the Department of Sociology, Indiana University, Bloomington, and are available for scholarly examination with permission of the author. Davis, Brody architects were interviewed by the author in New York on 13 March 1990; Cornell scientists and administrators were interviewed by the author (with the assistance of Mitch Berbrier and Rob Bienvenu, research assistants on the Lab Design Project) in Ithaca on 24 July 1990 and on 25–28 Septem-

ber 1990. Interview respondents have not been identified by name to insure their anonymity. All design documents were photocopied with permission from files in the office of the executive director of the Biotechnology Program at Cornell. The author gratefully acknowledges the willingness of Davis, Brody architects and Cornell scientists and administrators to allow a sociologist to examine freely every detail of the CUBB.

The hypermedia version of the Cornell Biotechnology Building was built by Peter C. Honebein, with assistance from Pai-Lin Chen and William Brescia. This project was initially a portion of the 1991 Sociological Research Practicum, and received funding from the Institute of Social Research at Indiana University. A description of the goals and technical means of the hypermedia project is available from the author. Interview transcripts were entered into the hypermedia environment as "text" files and searched directly via the HyperKRS Search Application. Other documents (the design paper trail, blueprints, photographs) were scanned in as "graphic" files, and the search was conducted through keywords electronically appended to each document.

15. Davis, Brody's "Program and Project Concept" phase report begins with the justification for a new biotechnology building at Cornell, situates the project within the organizational context of the division of Biological Sciences, lists members of the design team, and outlines the basic criteria said to guide design decisions. Each "functional unit" (e.g., "administration" or "research space" at the highest level of aggregation, "flexible environmental chamber" or "fly rooms" at the lowest) is described in terms of the activities to be done in it, the people who will occupy it, the desired square footage, the utilities requirements (electrical, mechanical, HVAC), and its relationship to other functional units (to resolve questions of "adjacencies" and "stacking"). These verbal and numerical descriptions of spaces are accompanied by plans showing the architecture of the proposed building—a two-dimensional graphic representation of spatial allocation of functions and utilities. As the project moves through schematics and design development, verbal and numerical representations are refined and revised, and the drawings assume sharper definition and detail.

16. Interestingly, in a memo written about seven weeks earlier, from the director of the Biotechnology Program (a scientist and administrator) to another university official (dated 12 April 1984), the "brief justification of the space, cost and site associated with the Biotechnology Building" makes no mention of "public" or "service," but instead describes a need to bring together in one place people and equipment in the "molecular aspects of genetics and cell biology." The memo says nothing about anticipated consequences for anyone outside Cornell—though it does mention an amazingly prescient (and hefty) price tag of $30 million, something that could not be ignored as later justifications for the building expanded the expected beneficiaries to include publics beyond the university.

17. A description of the "Site Program Requirements" prepared several months later (10 Dec. 1984) is even more explicit about the exclusion of this new public for (from) the building: "Corson-Mudd [an adjacent biology building] and Biotech are research facilities, and do not include teaching laboratories or classrooms. Undergraduate students are not normally expected to be in these buildings."

18. A Davis, Brody architect told me that Cornell did not want anyone "wandering through a laboratory, potentially being exposed to stuff that he doesn't even know he's being exposed to." The idea was incorporated into the conceptual phase report of 16 April 1985: "Due to the potential of accidental exposure to dangerous substances, the building design should clearly define research areas, with proper security and isolation."

19. Peter Galison, "The Trading Zone: Coordination Between Experiment and Theory in the Modern Laboratory," paper presented at the International Workshop on the Place of Knowledge, Tel Aviv and Jerusalem, May 1989; Galison, "Computer Simulations and the Trading Zone," in *The Disunity of Science: Boundaries, Contexts, and Power*, ed. Peter Galison and David J. Stump (Stanford, Calif.: Stanford University Press, 1996), pp. 118–57. The minutes of a building committee meeting one week later (24 Jan. 1985) read: "Visitors to be directed to contact points." As we shall see in the case of laboratories for "visiting corporate scientists," some visitors get further inside the CUBB—and stay there longer—than others.

20. One advantage in doing an electronic search for public and private is that the computer, with no a priori assumptions about the potential sociological interest of a given appearance of a defined character string, overlooks none. A search for *privacy* once took me to a report on telecommunications requirements. At first I was excited, hoping to find some juicy discussion of communications, but alas, the "privacy principle" in question did not belong to humans, but "applied to all Telecom (voice) wiring. The wiring is to be run in its own enclosures separated from all other wiring." A later memo faulted the architects for forgetting this privacy principle, as they crammed several wires into the same conduit!

21. Mario Cuomo, speech at the CUBB dedication ceremony, 15 May 1989. One scientist indicated that the public (taxpayers) and private (corporate) parts are not necessarily aligned: "Companies are giving us money so we could put up this building, and I assume if they are doing that, they figure . . . that it's worth it to them. Or they wouldn't be doing it anymore, giving us money. But, if you are talking about [whether it is worth it] to the *public*. . . I don't know."

22. Latour, *Science in Action* (note 5), p. 162.

23. The lobby has been used for weddings and ballroom dances as well as scientific meetings.

24. A faculty scientist in the eukaryotes research group told me: "The common policy here is that each faculty has the same space. Whether it is a junior faculty with five students or a senior faculty with twenty students and postdocs. That was a sort of general policy of this [building] committee."

25. The Executive Director of the Biotechnology Program told me that "people that had a large amount of space [before] the move . . . saw that as losing. But in Wing Hall [their previous home], we didn't have common equipment rooms, so people basically housed their own equipment. Well, if you had a lot of equipment of your own, and suddenly saw your space shrinking—there was some panic set in! But then, when they realized that the equipment would go in separate spaces, which wasn't counted as part of your [private] square footage, that smoothed out again."

26. A faculty scientist in the eukaryotes research group said: "The lab is essentially all the same, but the special rooms are designed according to who will be on that floor or on that side of the floor. For instance, here we need . . . a plant growth room and a plant tissue culture room, which is not needed by some other group. But they need maybe a space for flies, or they need some additional space for something else. . . . The big labs, they are all the same. The building architects and planners said we should not vary the basic design, which I think is reasonable. We didn't insist on changing that, but the small rooms yes, they are of different size, different electricity supply or temperature control, whatever it is we need, are supposed to be installed." The Director of the Biotechnology Program, and a faculty member in the cell membranes group, also suggested that "people did have the opportunity to put a lot of input into the space that they would use as common space."

27. Mainly because the technology of electrophoresis became cheaper, smaller, safer, and more often used, it was later decided to house it on wet benches in the

private labs, obviating the need for a special room in communal space. Is this a *failure* to colonize the communal? Perhaps. But it points out a tradeoff inherent in the strategy: successful colonization—locating equipment outside the lab—means more steps are needed (a loss of time so precious in the fast-paced world of biotechnology). On the pace of biotechnology, see Michael Fortun's article in this volume.

28. These and following passages are from a first draft of "Bio-Tech Space Program," 21 February 1985.

Index

Contributors

Robert Bud is Head of Life and Communications Technologies at the Science Museum London and is responsible for collections research at the museum. He is author of *The Uses of Life: A History of Biotechnology* (1993) and coeditor with Deborah Warner of *Instruments of Science: An Historical Encyclopedia* (1997). He is currently exploring the ways museum objects can inform the history of technology, working with Bernard Finn of the National Museum of American History and Helmuth Trischler of the Deutsches Museum.

Alberto Cambrosio is Associate Professor in the Department of Social Studies of Medicine at McGill University, where he teaches medical sociology with a special interest in the sociology of biomedical knowledge and practices. His work focuses on the material culture of late twentieth-century medicine. He is coauthor with Peter Keating of *Exquisite Specificity: The Monoclonal Antibody Revolution* (1995).

Angela N. H. Creager is Assistant Professor in the Department of History and Program in History of Science at Princeton University, where she has been teaching history of biology since 1994. She is currently working on a book-length account of the importance of virus research to the growth of postwar biology in the United States. Her other research interests are directed at understanding the impact of World War II on biomedical research and using comparative approaches to the history of molecular biology.

Michael Fortun is Executive Director of the Institute for Science and Interdisciplinary Studies (ISIS) at Hampshire College, Amherst, Massachusetts, a non-profit organization combining the pursuit of science with science studies scholarship and citizen involvement. Besides continuing his research on the science, industry, and culture of genomics, he is following and writing on recent work in theoretical and experimental quantum physics. He is coauthor with Herbert J. Bernstein of *Muddling Through: Pursuing Science, After the Fact* (forthcoming).

Thomas F. Gieryn is Professor of Sociology at Indiana University (Bloomington), and Director of its Program on Scientific Dimensions of Society. He is author of *Cultural Cartography of Science: Episodes of Boundary-Work, Sociologically Rendered* (forthcoming). He was the Ralph and Doris Hansmann Member for 1996–97 at the School of Social Science, Institute for Advanced Study, Princeton.

Herbert Gottweis is Associate Professor in the Department of Political Science, University of Salzburg, Austria. Prior to that he was Assistant Professor in the Department of Science, Technology, and Society at Cornell University. He is the author of a book on the culture of legislation and of numerous articles and book chapters on comparative energy, environmental, and technology policy, as well as *Governing Molecules: The Discursive Politics of Genetic Engineering in Europe* (forthcoming).

Stephen Hilgartner is Assistant Professor in the Department of Science and Technology Studies, Cornell University. His research focuses on the social study of contemporary biological sciences, and he is conducting a prospective ethnographic study of the social world of genome mapping and sequencing. His articles have appeared in such publications as *Science Communication*, *Knowledge*, *Social Studies of Science*, *Handbook of Science and Technology Studies*, and *American Journal of Sociology*.

Lily E. Kay is Associate Professor of the History of Science at the Program in Science, Technology, and Society at the Massachusetts Institute of Technology. Her research has focused on the history of molecular biology. She is the author of *The Molecular Vision of Life: Caltech, the Rockefeller Foundation, and the Rise of the New Biology* (1993), and *Who Wrote the Book of Life? A History of the Genetic Code* (forthcoming). She is currently working on the history of modern neuroscience.

Peter Keating is Associate Professor in the History Department of the Université du Québec à Montréal; he teaches the history of science and medicine in that department and in the university's Science, Technology, and Society program. He is the author of *La science du mal: L'institution de la psychiatrie au Québec* (1993) and, with Alberto Cambrosio, of *Exquisite Specificity: The Monoclonal Antibody Revolution* (1995).

Martin Kenney is Professor of Applied Behavioral Sciences at the University of California, Davis, where he teaches technology and society and regional economic development courses. He is the author of *Biotechnology: The University-Industrial Complex* (1986), *The Breakthrough Illusion* (1990), and *Agents of Innovation: Venture Capital and High Technology* (forthcoming). He has published over fifty journal articles and book chapters.

Daniel J. Kevles is Koepfli Professor of Humanities and head of the Program in Science, Ethics, and Public Policy at the California Institute of Technology. His works include *In the Name of Eugenics: Genetics and the Uses of Human Heredity* and *The Physicists: The History of a Scientific Community in Modern America* as well as numerous articles and reviews on issues in science and society past and present, including the engineering and ownership of life. He is currently completing a book on the Baltimore case.

Sheldon Krimsky is Professor in the Department of Urban and Environmental Policy at Tufts University. He trained in physics, philosophy, and philosophy of science, and his research has focused on the social and ethical impacts of science and technology. His is the author of *Genetic Alchemy* (1982) and *Biotechnics and Society* (1991), coauthor of *Environmental Hazards: Communicating Risks as a Social Process* (1988) and *Agricultural Biotechnology and the Environment* (1996), and coeditor of *Social Theories of Risk* (1992). He is completing a two-year study of the "environmental endocrine hypothesis" and public policy.

Arnold Thackray is President of the Chemical Heritage Foundation. Educated in England, he was a fellow of Churchill College, Cambridge. He served on the faculty of the University of Pennsylvania for over a quarter of a century and was the founding chairman of, and most recently Joseph Priestley Professor in, the Department of History and Sociology of Science. His publications include *Gentlemen of Science* (1981) and *Chemistry in America* (1985).

Susan Wright is Head of the Science Program at the Residential College of the University of Michigan, where she teaches the history of science and international relations. She is the author of *Molecular Politics: Developing American and British Regulatory Policy for Genetic Engineering, 1972–1982* (1994) and coauthor of *Preventing a Biological Arms Race* (1990). Her current research addresses the north-south dimensions of the 1972 Biological Weapons Convention.

Acknowledgments

This book is one fruit of the continuing biomolecular sciences initiative (BIMOSI) of the Chemical Heritage Foundation's Beckman Center for the History of Chemistry.

The emerging "invisible college" of academic scholars studying the novel phenomenon of biotechnology is small and widely scattered. That dispersal led me, working with M. Susan Lindee of the Department of History and Sociology of Science at the University of Pennsylvania, to plan a way to draw those scholars together, to review the present state of our knowledge, and to establish some of the concepts, themes, and key episodes that are becoming central to our understanding of biotechnology and the rise of the biomolecular sciences.

Thanks to the excellent staff work of the Beckman Center and especially of Elizabeth Hanson, a graduate student at Penn, some eighteen scholars from points as far apart as Paris, France and Pasadena, California duly assembled at the Eagle Lodge Conference Center, just outside Philadelphia, from 28 to 30 October 1993. In that agreeable setting, the invisible college came alive. Though many of the individuals had not previously met, and though their nationalities and disciplines were varied, a common basis of knowledge, enthusiasm for the task, and excitement of the pioneer ignited the group from the very first moments. Precirculated papers meant that conference sessions were devoted to discussion, argumentation, and rethinking rather than to exposition. The informality of Eagle Lodge and generous free time between sessions increased the ability of individual scholars to enhance each other's understanding.

Following the conference, drafts were revised and subjected to the critique of anonymous referees. The final result is a volume of unusual coherence devoted to a subject of unusual and growing importance, an example of the best of modern history and sociology of science.

Especial thanks are owed to Susan Lindee and to Jeffrey L. Sturchio for many helpful suggestions, to Irene I. Lukoff for overseeing the logistics of the Eagle Lodge Conference, to Norton Zinder of Rockefeller University for bringing the perspective of the laboratory scientist to the conference, to Frances Coulborn Kohler for her meticulous skill in fine editing and producing this volume, and to Harriet Zuckerman and the Andrew W. Mellon Foundation for their enlightened support of the study of the modern biomolecular sciences.